Home Plumbing

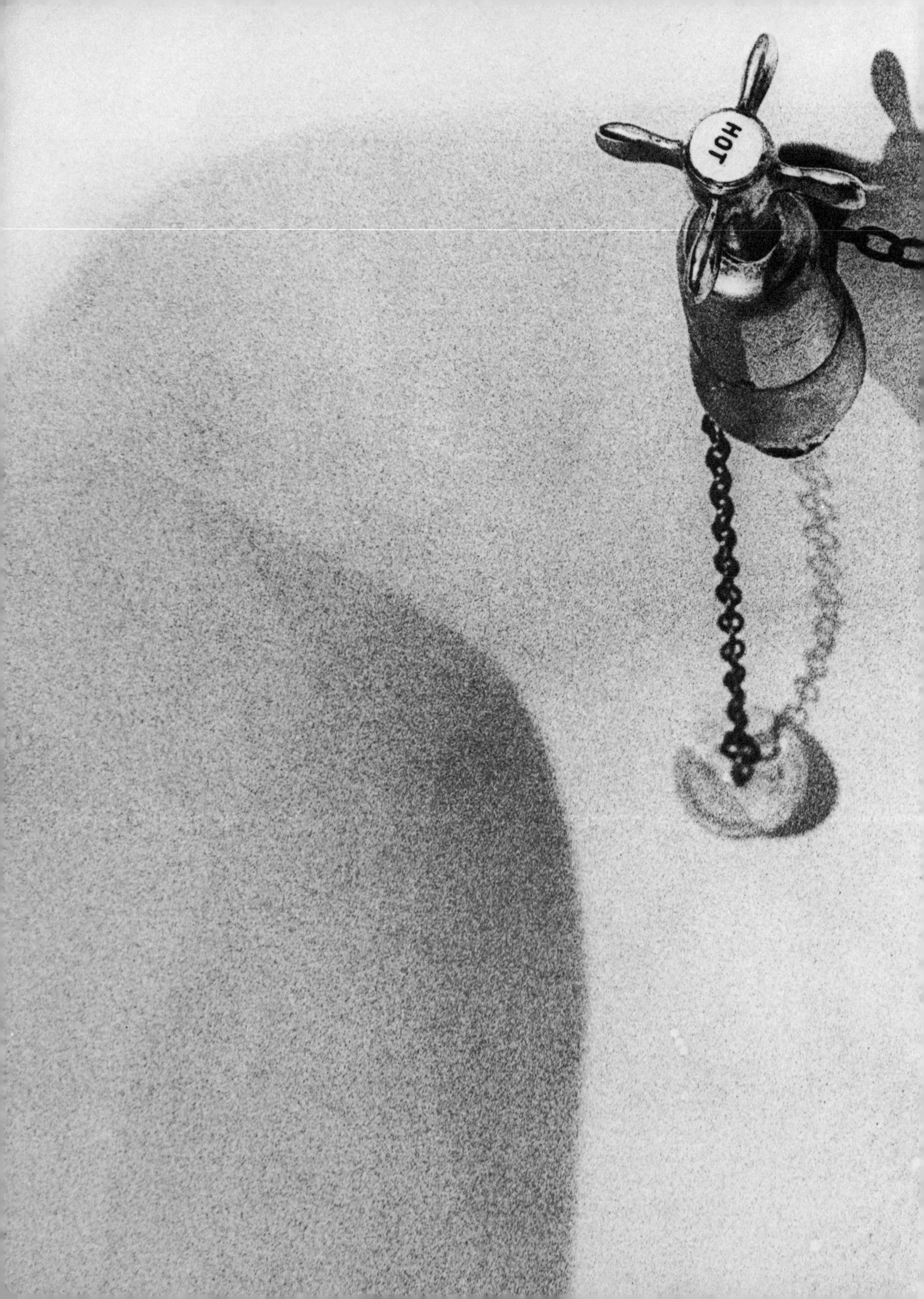
HOT

Home Plumbing

Ernest Hall FRSH

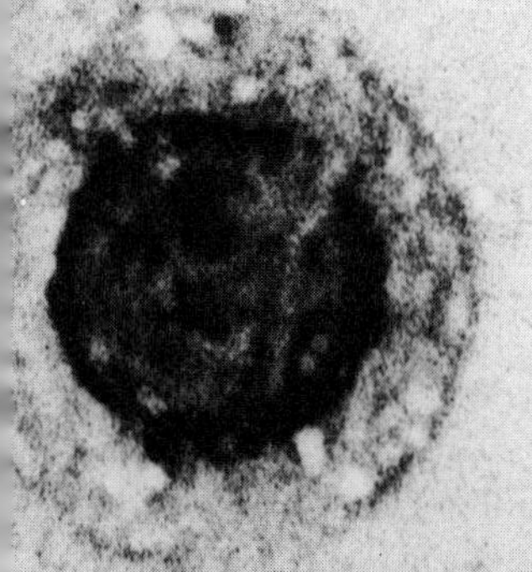

Newnes Technical Books

The Butterworth Group

UNITED KINGDOM

Butterworth & Co (Publishers) Ltd
London: 88 Kingsway, WC2B 6AB

AUSTRALIA

Butterworths Pty Ltd
Sydney: 586 Pacific Highway, NSW 2067
Also at Melbourne, Brisbane, Adelaide
and Perth

CANADA

Butterworth & Co (Canada) Ltd
Toronto: 2265 Midland Avenue, Scarborough,
Ontario, M1P 4S1

NEW ZEALAND

Butterworths of New Zealand Ltd
Wellington: 26–28 Waring Taylor Street, 1

SOUTH AFRICA

Butterworth & Co (South Africa) (Pty) Ltd
Durban: 152–154 Gale Street

USA

Butterworth (Publishers) Inc
Boston: 19 Cummings Park, Woburn, Mass 01801, USA

First published 1977 by Newnes Technical Books
a Butterworth imprint

ISBN 0 408 00246 8

Printed in England by Butler & Tanner Ltd. Frome and London

Preface

It is the ambition of most of us to have a place we can call our own. This is commendable, but it does not take long to discover that privilege brings responsbilities. A home will not look after itself, and as the occupant you must be prepared to do battle with damp, rust and corrosion, fading, peeling, wormholes, leaks, blown fuses—and much more besides.

Of course you can call in a man, but with so much of the work labour-intensive—the areas where costs are now so very high—you will have to delve deep in your pocket to keep up with the bills. From this stems the great incentive to tackle the work yourself when you can reduce costs to materials only.

The term do-it-yourself encompasses a very wide field of activity, and there is much to learn. It is not always easy, but once new skills have been mastered d-i-y becomes rewarding and satisfying. The books in this new series which as well as plumbing cover subjects ranging from home decoration, repairs, electric wiring and heating, are designed to help you acquire the necessary skills. All you need to add is practice!

They are written by people with very considerable practical experience in the d-i-y field, and all have been involved in feature-writing for DIY magazine over the years. The authors have also been responsible for dealing with hundreds of readers' queries—which has given them an invaluable insight into the problems encountered in and about the house.

I'm sure you will find their advice invaluable. May I wish you success in all you undertake.

Tony Wilkins
Editor, 'Do it Yourself' Magazine

Acknowledgements

I gratefully acknowledge the help and encouragement that I have received in the preparation of this book from the following firms and organisations and also from my friend and colleague, Charles Burley, who read through the typescript in the light of his 45 years practical experience in the building trade.

Armitage Shanks Ltd., Armitage
Barking Brassware Co Ltd., Barking
Conex Sambra Ltd., Tipton
Deltaflow Ltd., Crawley
Do-it-yourself Editorial Staff
The Electricity Council
Fordham Pressings Ltd., Wolverhampton
Glynwed, Bathroom and Kitchen Products Ltd., Long Eaton
Ideal Standard Ltd., Hull
IMI Range Ltd., Stalybridge
Ingol Precast Ltd., Preston
Kay & Co Ltd., Bolton
Key Terrain Ltd., Maidstone
Marley Extrusions Ltd., Lenham
Osma Plastics Ltd., Hayes
Peglers Ltd., Doncaster
Rokcrete Ltd., Clacton-on-sea
Sofnol Ltd., Thaxted
Twyfords Ltd., Stoke-on-Trent
Walker Crossweller Ltd., Cheltenham
Wednesbury Tubes Ltd., Bilston

Note on metrication

The progress of metrication has not, at the time of writing, reached completion as far as plumbing equipment is concerned.

I suspect that, for many years after this process has been completed, a great many British householders—and plumbers—will continue to ***think*** in terms of feet, inches and gallons rather than in terms of metres, millimetres and litres.

Because of this I have, throughout this book, attempted to supply both metric and imperial measurements where both seemed to be of value. The figure given in brackets, whether metric or imperial is the one that at the time of writing seemed to me to be of lesser importance.

It should be noted that the metric ***equivalent*** is not necessarily a straight translation of the imperial value. For instance the 15mm metric copper tube is the equivalent of the ½ in imperial copper tube—but 15mm does not equal ½ in. The reason for this apparent disparity is that the imperial measurement is of the internal diameter of the tube, the metric measurement is of the external diameter.

Ernest Hall

Contents

Introduction
The scope of home plumbing

Plumbing has not quite the same meaning for the tradesman plumber as it has for the householder. To the man in the trade it means the ability to work skilfully in lead, zinc and copper. It means knowledge and experience of solders and fluxes, of stocks and dies, soakers and flashing, red lead and putty, tank shears and Stillson wrenches and pipe bending machines.

To the householder, for whom 'Home Plumbing' is intended, ***plumbing*** means his domestic water services. It means the hot and cold water taps, the hot water system, the sink, the wash basin and the bath. It means the lavatory and the underground drains. The householder's interest in plumbing begins with the service pipe bringing water into his home from the main. It ends with the final length of drain connecting the the manhole inside his front garden with the public sewer.

It is this difference in outlook that makes the tradesman plumber suspicious of the

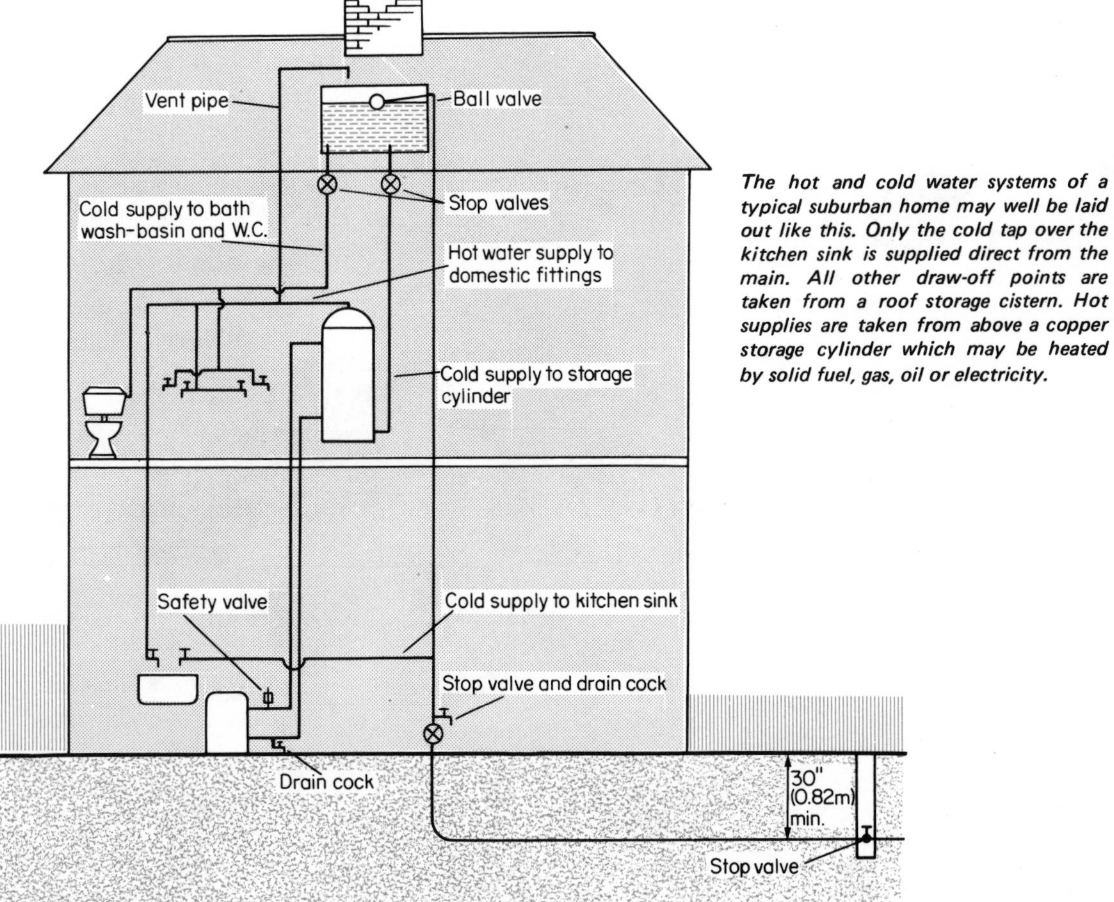

The hot and cold water systems of a typical suburban home may well be laid out like this. Only the cold tap over the kitchen sink is supplied direct from the main. All other draw-off points are taken from a roof storage cistern. Hot supplies are taken from above a copper storage cylinder which may be heated by solid fuel, gas, oil or electricity.

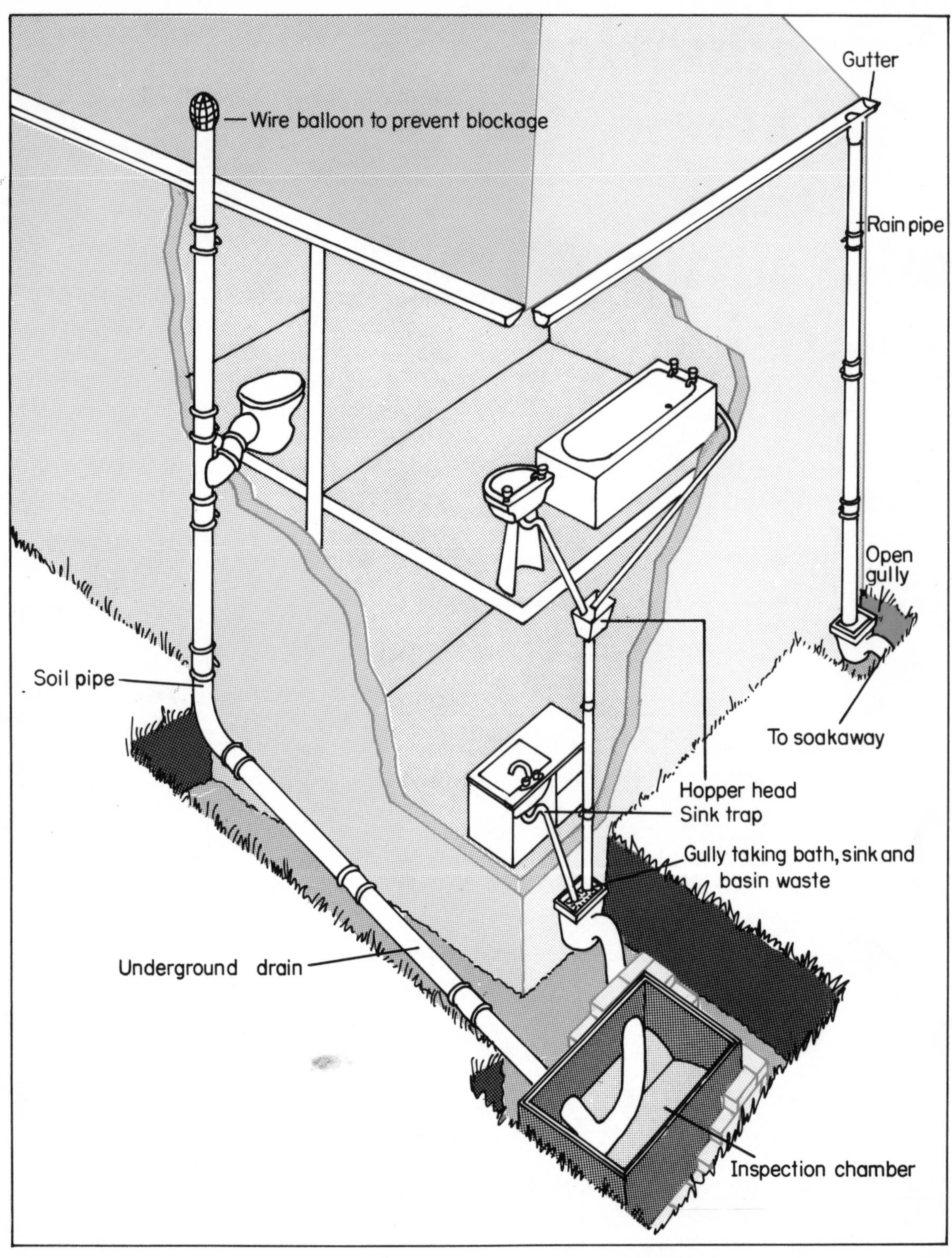

A house built over a decade ago will probably have a 'two pipe' drainage system. 'Soil fittings' w.c.s and urinals—were connected directly to the drain via a main soil and vent pipe. 'Waste fittings'—baths, basins, bidets and sinks—discharged over trapped yard gullies.

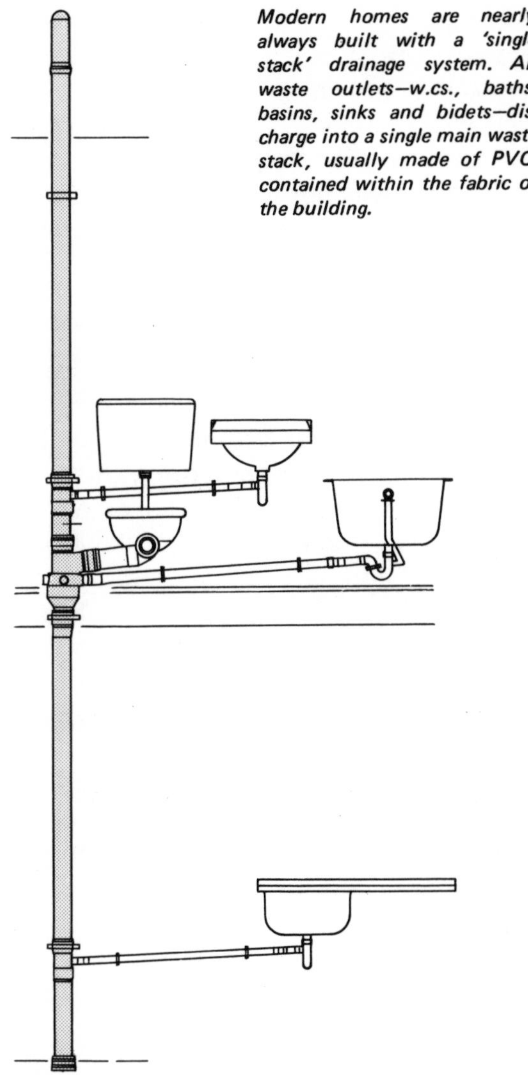

Modern homes are nearly always built with a 'single stack' drainage system. All waste outlets—w.cs., baths, basins, sinks and bidets—discharge into a single main waste stack, usually made of PVC, contained within the fabric of the building.

do-it-yourself enthusiast. How can a householder—a bank or insurance official, a civil servant, a doctor or a lawyer—possibly acquire, in his spare time, skills to which the tradesman has devoted a life's work?

The householder neither wants, nor needs, to acquire these skills. He does want to know how his domestic plumbing system works and how to protect it from frost and corrosion, how to identify faults and how to carry out emergency repairs. How many professional plumbers welcome an urgent call to clear a sink waste pipe or to renew the washer on a tap?

Over a decade of replying to plumbing queries sent in to the readers' problems service of Do-it-Yourself magazine; has taught me that intelligent and determined householders can do far more than simply carry out emergency repairs. Readers have written to tell me how, using the materials and techniques now available to the handyman, they have converted their lavatory suites from high to low level, have renewed baths, sinks and wash basins and have installed complete hot water and central heating systems. For d.i.y. enthusiasts planning projects of this kind, this book will prove to be of value.

I hope too, that it will also interest the householder who is unlikely to attempt on his own accord anything more ambitious than curing a dripping tap or overflow pipe; the man who would like to stop the ball-valve of his cold water storage tank making *quite* so much noise in the middle of the night, or who wishes that he had known what was involved when the plumber shook his head solemnly and said, 'you'll really have to have an indirect system, sir'.

I would like to prove to him that his plumbing system need not be the noisy, metallic monster of his imagination, threatening simultaneously to engulf his possessions in scalding—or icy—water, and his family in debt.

Finally, as a former local authority Public Health Inspector and Housing Manager, I hope that this book may be of some value to trainees in these and similar professions who need to be thoroughly familiar with domestic plumbing and drainage but who do not need to acquire many of the professional plumber's practical skills.

An efficient, unobtrusive plumbing system is the mark of a really comfortable home. This book will tell you how your plumbing system measures up to modern standards, and how easy it would be to bring it up to date.

Chapter 1
Cold water services and the cold water storage cistern

Home plumbing begins with the service pipe bringing water into the home from the main. The householder's responsibility for this pipe extends from the Water Authority's stop-cock which will be in a purpose-made pit in the footpath or roadway. This will probably have a specially shaped shank which can be turned only by means of one of the Water Authority's turn-keys.

Pre 1939 the service pipe would have been of lead. In a more recently built home it will be of 15 mm (½ in) copper tubing or, just possibly, of polythene tubing. It must be at least 0.82 m (2 ft 6 in) below the surface of the ground and, in order to permit any air in the pipe to escape, it should rise slightly towards the house.

It is very important, as a frost precaution, that the minimum depth of 0.82 m is maintained throughout the length of the pipe. It sometimes happens that an enthusiastic landscape gardener creates a sunken garden above the service pipe, reducing its effective depth by half. This could be disastrous during a period of severe frost.

Where the service pipe is taken through the foundations of the house it should be threaded through a length of drain pipe to protect it from being crushed as a result of any slight settlement that may take place. The kitchen floor, through which the pipe will rise into the house, is most likely to be of solid construction. If the service pipe has to enter the house through the open underfloor space of a boarded floor, it should be very carefully lagged against frost.

One way to do this is to take it up through the centre of a 100 mm (4 in) or 150 mm (6 in) drain pipe and to fill the space between the service pipe and the sides of the drain pipe with vermiculite chips.

The service pipe—now often called the 'rising main'—should rise into the house against an internal wall in the kitchen. Immediately above the floor should be the householder's own stop-cock with, just above it, a drain-cock.

These two fittings enable the water supply to the house to be cut off and the rising main drained when required. You should make sure that every member of the household knows where this stop-cock is situated

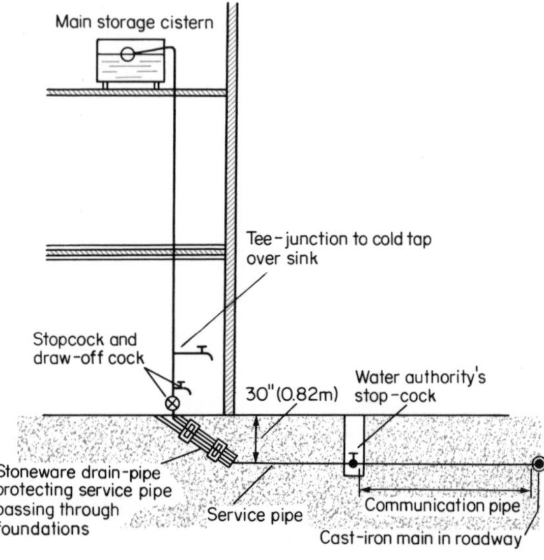

As a frost precaution the service pipe bringing water into the home from the water authority's main must be at least 0.82 mm (2 ft 6 in) below ground level throughout its length. It should always be taken up to the main storage cistern against an internal wall of the house.

Where the service pipe or 'rising main' enters the house through a hollow boarded floor, special precautions are necessary to protect it from icy underfloor draughts. It is best threaded through a 150 mm (6 in) stoneware drain pipe packed with vermiculite chips or similar insulating material.

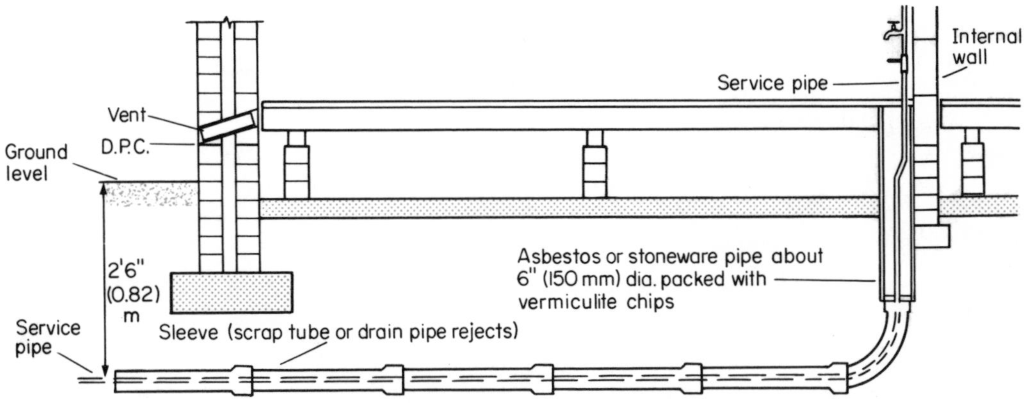

and how to use it. In an emergency—a burst pipe, a leaking cold water storage tank or a jammed ball valve—turning off this stop-cock will immediately stop any further flow of water into the house and will limit any damage which might occur.

It is a good idea too, to turn this stop-cock on and off two or three times at intervals of six months or so. A stop-cock long disused, can jam and prove to be useless when most needed.

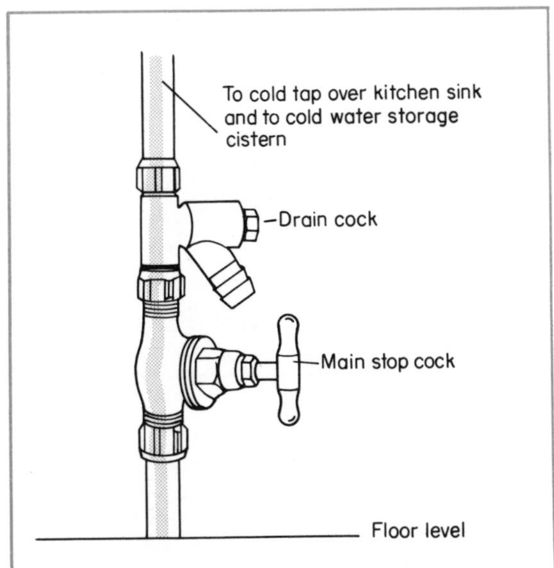

The householder's main stop-cock which is often to be found under the kitchen sink, enables the water supply into the house to be cut off at will. The drain-cock immediately above it makes it possible to drain the rising main.

From this stop-cock and drain-cock, the pipe will rise vertically to discharge by means of a ball valve into the main cold water storage tank or cistern. In some households, the lavatory flushing cistern and the bathroom cold taps will be supplied direct from the rising main. Many Water Authorities however require these draw-off points to be supplied from a storage tank. Only the cold tap over the kitchen sink—supplying water for drinking and cooking—and perhaps a garden supply, are permitted to be taken direct from the main.

Cold water storage cistern

The main cold water storage cistern probably causes more anxiety to the householder than any other piece of plumbing equipment. It is out of sight, and usually in a spot where it cannot be readily inspected. If it leaks or overflows, the resultant flood may do hundreds of pounds worth of damage to ceilings, furnishings and carpets.

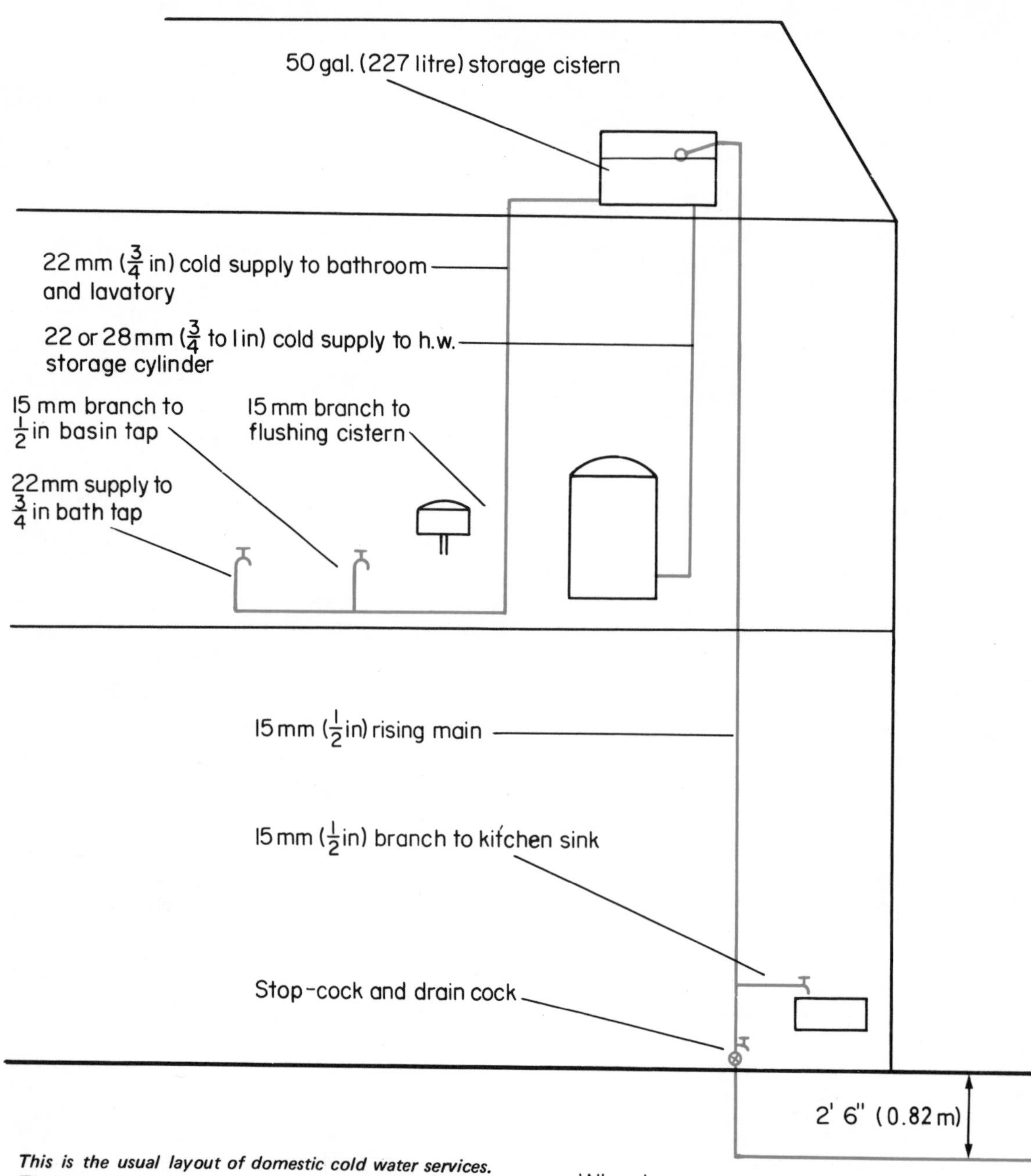

This is the usual layout of domestic cold water services. The only connections made directly to the rising main are the cold tap over the kitchen sink and the ball-valve connection to the cold water storage cistern. With the permission of the Water Authority a garden or garage supply might also be taken from the rising main. No branch cold supply pipe must be taken from the 22 mm or 28 mm (¾ in or 1 in) cold supply pipe from the cold water storage cistern to the hot water cylinder. Distribution pipes are connected to the storage cistern at points 2 in above the cistern's base to reduce the risk of grit or debris being drawn into the plumbing system.

Why have one at all? It is technically possible to connect all cold water services direct to the rising main. There are also gas and electric hot water appliances designed for mains connection.

However most Water Authorities require the provision of a substantial storage cistern in each home to act as a buffer between themselves and consumers at times of peak demand.

The storage cistern has advantages for the householder too. It provides a substantial reserve of water against breakdown of the mains supply. It also provides a supply of water at constant, relatively low, pressure to feed hot water supply apparatus. Most hot water systems demand a storage cistern. Although it is sometimes possible to incorporate such a cistern in the hot water system itself, it is generally more convenient to replenish the hot water system from a main cold water storage tank.

Where should the main cold water storage cistern be situated? The traditional site, up in the roof space, has come in for a good deal of criticism in recent years. It is argued that, by keeping the cistern out of the roof space it is more accessible for inspection, the risk of contamination is reduced and, above all, there is far less risk of frost damage to the cistern itself or to the pipes connected to it.

All this is perfectly true. Yet if the cistern is brought out of the roof space it must be placed in the upper part of an airing cupboard or in a special cupboard in the bathroom or bedroom. Here, since no cistern is entirely silent, it will make its presence known by its noise; and it will attract condensation.

Its relatively low level will mean a poor flow of water from bath and basin taps and a slow refill to the lavatory flushing cistern. It will make it much more difficult and expensive to install a shower.

On balance I feel that the roof space is still the best position for the storage cistern. It should be situated against a chimney breast taking a flue in constant use, lengths of pipe in the roof space should be kept as short as possible and these pipes, and the cistern itself, should be thoroughly protected against frost. Methods of doing this will be discussed in a later chapter.

Wherever the cistern is situated it must be properly supported, preferably above one of the dividing walls of the house. A gallon of water weighs 10 lb so that a cistern with a capacity of 50 gal (227 litres) will contain over 4 cwt of water quite apart from the weight of the cistern itself.

Most Water Authorities require that storage cisterns of this kind should have an ***actual*** capacity of 50 gal (227 litres). This is the capacity to a water level 112mm (4½ in) from the cistern's rim.

Galvanised steel storage cisterns

Galvanised mild steel is the traditional material of which storage cisterns are made. Tens of thousands of cisterns of this material are in use and are giving trouble-free service.

They have disadvantages though. They are heavy and generally need two men to manhandle them up into the roof space. Cutting the holes for pipe tappings is best done with special tools made for the purpose. The biggest disadvantage of galvanised steel cisterns, however, is their liability to corrosion. This drawback has increased with the, nowadays, almost universal use of copper tubing for plumbing.

It is well known that if connecting rods of zinc and copper are immersed in a weak acid—an electrolyte—the conditions of a simple electric cell are produced. Electric current will pass from one rod to the other, bubbles will form in the electrolyte and the zinc will dissolve away.

Something like this may happen when copper tubing is connected to a galvanised steel storage cistern. The water in the cistern will, if slightly acid, act as the electrolyte. The zinc coating of the galvanised steel may dissolve away and permit water to attack the steel underneath. This process is called electrolytic corrosion.

Asbestos cement cisterns

Asbestos cement cisterns cannot corrode. They have rounded internal angles and smooth jointless walls which make for easy cleaning. Once installed and protected from frost they should last forever.

These cisterns are rather heavy. A typical asbestos cement cistern with an actual capacity of 50 gal weighs 104 lb. They are also liable to damage both during installation and from frost. They must be handled with care. Holes should not be bored nearer than 4 in to the base of the cistern. Tappings should be sealed off with two washers on each side of the cistern wall. One of these washers, the one against the cistern wall, should be of soft material.

As the cistern walls are ½ in thick, making holes for tappings can present difficulties. One manufacturer recommends the following procedure: Mark out the circumference of the hole and drill a complete circle of small holes inside this circumference. Use an ordinary brace and bit but with the bit ground to an angle of 20° instead of the usual 59°. When all the holes have been drilled, the piece in the centre can be pushed out and the hole finished with a half-round rasp.

This technique can be adopted by those with a minimal tool kit for cutting holes of any size in any storage cistern, tank or cylinder.

Plastic cisterns

Cisterns of plastic materials have advantages over both galvanised steel and asbestos cement. As well as being proof against corrosion, they are light, tough and easily fitted.

Plastic cisterns may be rectangular or circular in shape. The black polythene circular cisterns have the advantage that they can be flexed to pass through a relatively small trap-door into the roof space. A 50 gal capacity circular cistern will be 3 ft in diameter and almost 2 ft high. It can however be flexed to pass through any opening 2 ft square.

Plastic cisterns must always be supported on a flat, level platform. A piece of chipboard spiked to the rafters will meet this requirement.

Since they do not offer the same support as steel or asbestos cisterns, the rising main, when connected to a plastic cistern, must be firmly secured to the roof timbers. All pipes connected to such a cistern must join it squarely so as not to strain the cistern walls. Soft, plastic, washers must be used in direct contact with the cistern walls and no boss white or other sealing material

The makers of one brand of asbestos cement cistern recommend that holes for tappings should be made in this way. The householder with a minimal tool kit can use the same method for making circular holes in any material.

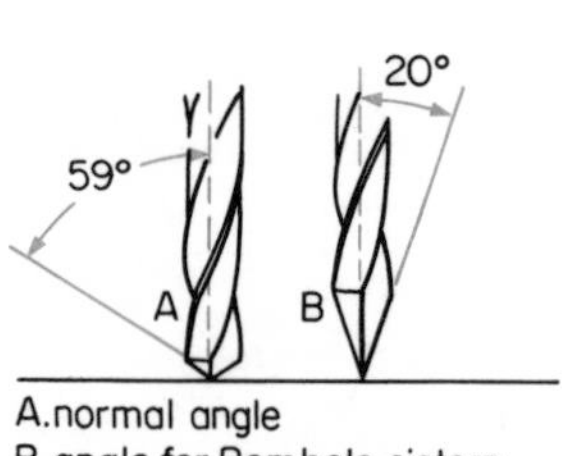

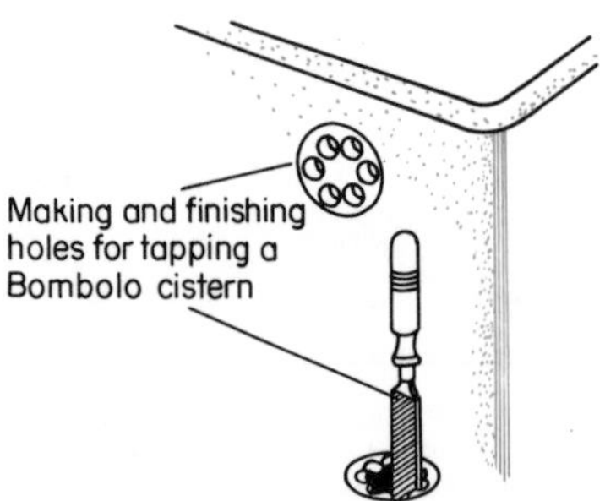

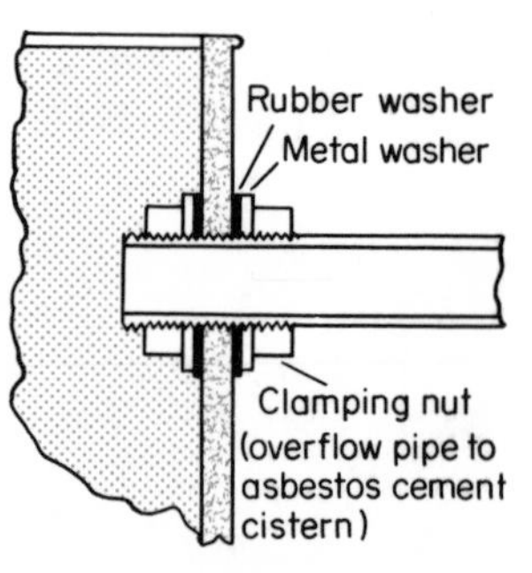

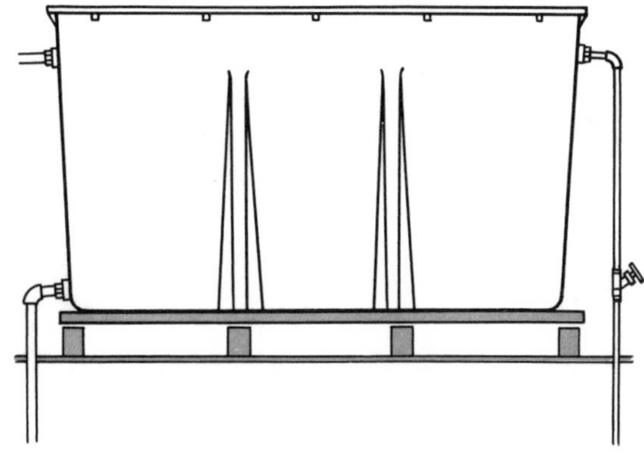

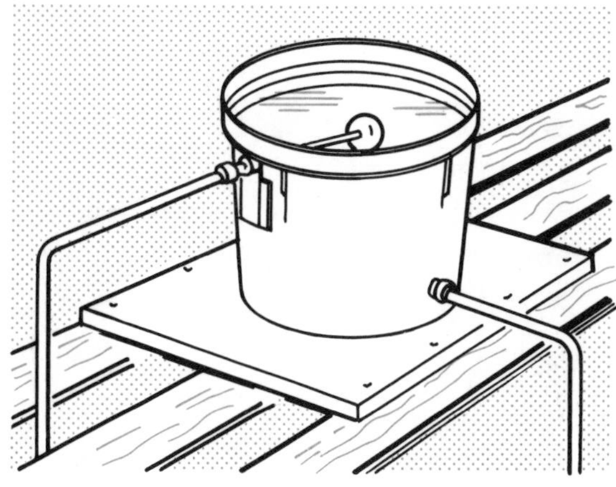

Plastic cisterns may be rectangular or round. They must rest on a flat, level base – not just on the ceiling joists – and care must be taken to ensure that all pipes connect squarely to the cistern walls. The great advantages of plastic cisterns are their lightness and their immunity to corrosion.

should be allowed to come into contact with the plastic.

Irrespective of the material from which the cistern is made, it should be provided with a dust-proof, but not air-tight, cover. Makers of asbestos cement and plastic cisterns often manufacture purpose-made covers that can be bought as an extra. It is however perfectly easy to make a lid of hardboard, plywood or asbestos board cut to size and provided with a 25 mm (1 in) wood strip fastened round its edges.

Connecting pipes

All cold water storage cisterns are supplied with water through a ball valve, usually fitted 37 mm (1½ in) below the cistern rim. The overflow or warning pipe must be fitted below the level of the ball valve inlet and about 25 mm (1 in) above the full water level of the cistern. This pipe will be a minimum of 22 mm (¾ in) in diameter.

There will normally be at least two pipes connected to the lower part of the cistern: a 22 mm (¾ in) diameter pipe supplying the bath cold tap with 15 mm (½ in) branches taken from it to the wash basin and the lavatory flushing cistern, and another 22 mm (¾ in) or 28 mm (1 in) pipe supplying cold water to the hot water storage cylinder.

These pipes should be connected to the cistern at a point at least 50 mm (2 in) above its base to reduce the risk of sediment from the mains being drawn into the pipes.

Faults in the cold water system

1. *Poor flow or poor pressure through draw-off points from the main* (cold tap over kitchen sink or ball-valve to storage cistern):

   Check that the main stop-cock is fully open.

   Check that the tap and ball-valve are functioning properly. See taps and ball-valves (Chapter 4).

2. ***Poor flow or poor pressure through draw-off points from storage cistern*** (bathroom cold taps or ball-valve to lavatory cistern):

   Check that taps and ball-valve are functioning properly and that a low pressure ball-valve is installed in the lavatory cistern.

   Try treatment suggested for air-locks in 'Faults in Hot Water Systems'.

3. ***Corrosion in cold water storage cistern*** (evident as a 'dusting' of rust on cistern walls, rust patches, particularly round tappings or warty outgrowths of rust and scale):

Drain and dry cistern thoroughly. Remove every trace of rust by wire brushing (use goggles to protect the eyes) or abrasive paper. Fill in any deep pit marks left by this process with an epoxy resin filler. Apply two coats of a ***tasteless and odourless*** bituminous paint. This treatment will give protection for two or three years and can be repeated as often as required.

A galvanised steel cistern not yet showing signs of rust can be protected from corrosion by means of a sacrificial anode. This is a block of magnesium immersed in the water and in electrical contact with the cistern walls. Magnesium has a high potential and electrolytic action will take place between the magnesium and the zinc coating of the cistern—to the advantage of the zinc. The magnesium block will slowly dissolve away—will be sacrificed—and the galvanised steel protected. This method has proved to be most effective in hard water areas.

4. ***A leaking cold water storage cistern*** (first indication may be water dripping through the ceiling of room below):

Immediately turn off main stop-cock and open up bathroom taps. This will drain the cistern and limit the damage.

Only after doing this should you climb into the roof space to investigate and to mop up between the rafters.

A leaking cistern will generally need replacement but the trouble could be due to a jammed or otherwise faulty ball-valve. See 'ball-valves'.

5. ***Water hammer*** (heavy drumming noise in the pipes, especially when a tap is turned off or on):

Usually due to a faulty tap or inefficient ball-valve. See 'taps' and 'ball-valves'.

Chapter 2
Domestic hot water supply-cylinder storage systems

It is difficult to recall that less than three decades ago the hot water system in most British homes consisted of a kettle on a gas ring, supplemented perhaps by a temperamental 'geyser' in the bathroom and a solid fuel, gas or electric clothes boiler.

Nowadays an efficient supply of hot water on tap is regarded as essential in every home. This has been recognised by successive governments who have included a hot water system among the basic amenities towards the installation of which any home owner can claim, as a right, a cash grant from his local Council.

The cylinder storage system of hot water supply is one of the most versatile and popular means of obtaining domestic hot water on tap. Originally used always in conjunction with a solid fuel boiler, perhaps supplemented by an electric immersion heater in the summer, it can be used with any fuel and can be adapted both to supply hot water and to provide a central heating system.

A simple 'direct cylinder' system is illustrated. The cold water storage cistern, hot water storage cylinder and boiler are ideally situated in a vertical column. This arrangement cuts down lengths of pipework and also means that any waste heat from boiler and cylinder rises to give the cold water cistern a measure of protection against frost.

The cold water supply to the cylinder is taken from a point 50 mm (2 in) above the base of the cold water storage cistern to a tapping near the base of the cylinder by means of a supply pipe at least 22 mm (¾ in) in diameter. The flow pipe from the boiler—probably 28 mm (1 in) in diameter—is taken from the upper tapping of the boiler to the higher of two tappings provided in the cylinder wall. From the lower tapping in this wall another 28 mm (1 in) return pipe is taken to the lower or return tapping of the boiler.

A 22 mm (¾ in) vent pipe rises from the apex of the cylinder dome to terminate open-ended over the cold water storage cistern. From this vent pipe is taken the 15 mm (½ in) hot water supply to the kitchen sink and a 22 mm (¾ in) supply pipe to the hot tap of the bath. From the latter pipe line a 15 mm (½ in) branch will be taken to supply the hot tap of the bathroom wash basin.

Since the hot water supplies to the kitchen and bathroom taps are taken from ***above*** the storage cylinder it will be obvious that the cylinder, boiler and flow and return pipes, cannot be drained from these taps. To enable the whole system to be drained when required, a drain-cock must be provided, close to the boiler, on the return pipe from cylinder to boiler. In addition, a spring-operated safety valve is often fitted close to the boiler. Traditionally this valve is fitted into the flow pipe from boiler to cylinder, though there is something to be said for locating it on the return pipe adjacent to the drain cock.

Most direct cylinders intended for use with solid fuel boilers are provided with an immersion heater boss in the dome. Into this can be screwed a long, vertically fixed immersion heater to provide hot water

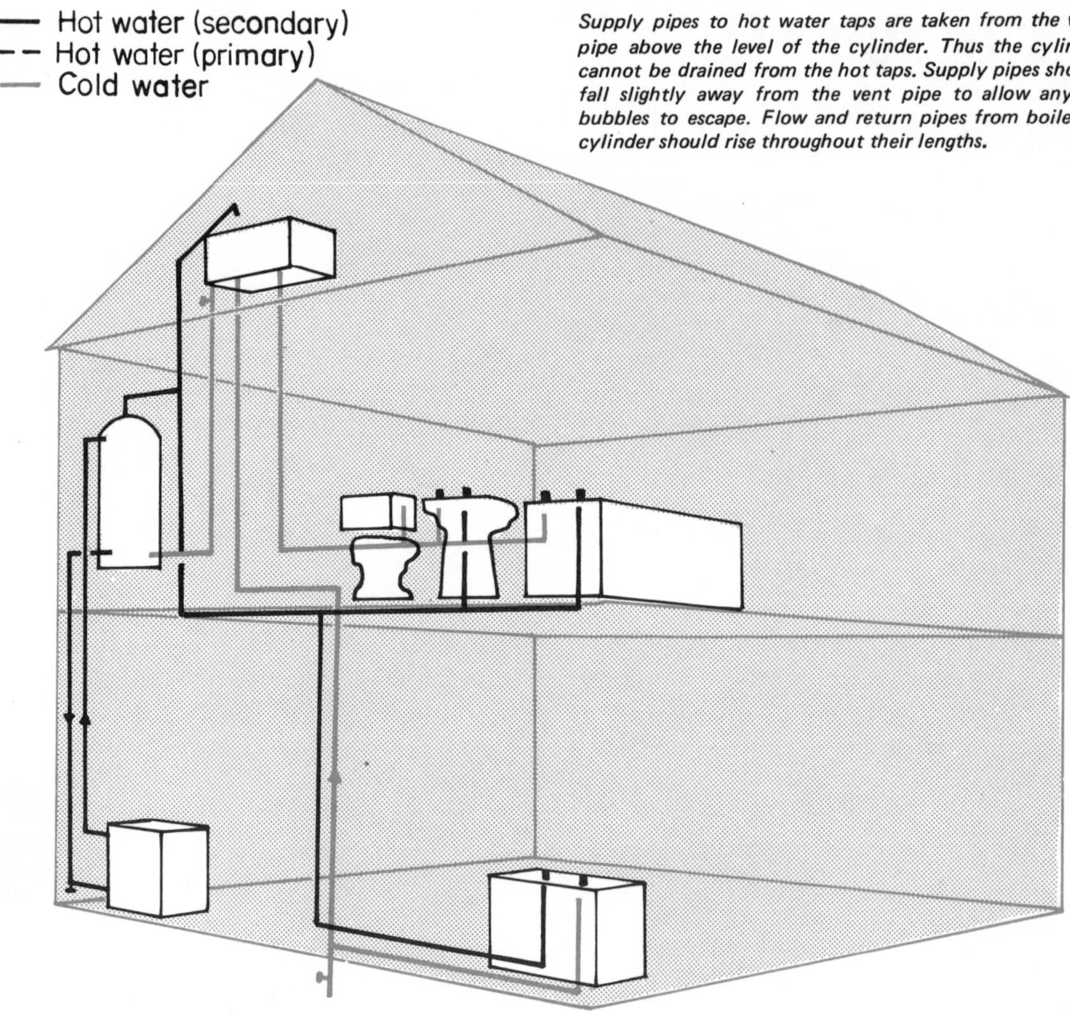

Supply pipes to hot water taps are taken from the vent pipe above the level of the cylinder. Thus the cylinder cannot be drained from the hot taps. Supply pipes should fall slightly away from the vent pipe to allow any air bubbles to escape. Flow and return pipes from boiler to cylinder should rise throughout their lengths.

during the summer months when the boiler is not in operation.

This is how a simple cylinder storage system works:

The boiler fire, when lit, heats the water in the boiler. As water is heated it expands and pint for pint—or litre for litre—weighs less than it did when cold. Colder, denser and heavier water from the return pipe then flows into the boiler pushing the warmer, lighter water up the flow pipe into the cylinder. In other, rather less accurate but more familiar, words 'hot water rises' and is replaced by cold. Circulation has begun and will continue for so long as the boiler fire is alight.

The warm water enters the cylinder near its dome and, since it is lighter in weight that the other water in the cylinder, it will 'float on top of it', remaining at the top of the cylinder and gradually extending downwards as circulation continues.

Since the supply pipes to the hot taps are taken from above the cylinder it will always be the hottest stored water, from the upper part of the cylinder, that will be drawn off.

As water is drawn off from the hot taps, cold water will flow in to the lower part of the cylinder from the cold water storage cistern. This will, in its turn, pass down the return pipe to be heated in the boiler.

The demand for the speedy installation of compact hot water systems into homes that had previously lacked them, and into flats converted from older, larger houses, resulted in the production of packaged hot water systems. Some of these are advertised, with justification, as being complete 'packaged plumbing systems'.

The earliest on the market consisted of a copper hot water storage cylinder, usually of 25 gal capacity, with a small feed cistern—also cylindrical and made of copper, immediately above it. The small feed cylinder was sufficiently large to supply the hot water system, but not to provide a cold water supply to the bathroom and lavatory.

Systems of this kind could therefore only be installed in those areas where the Water Authority permitted bathroom and lavatory cold water draw-off points to be taken direct from the main.

(a) a small packaged or 'two-in-one' hot water system. The small cold water storage cistern of such a unit would be capable of supplying the hot water cylinder only. Supplies to all cold taps and flushing cisterns would have to be taken direct from the rising main.

(b) a complete 'packaged plumbing system' with full size 50 gal cold water storage cistern and 25 gal hot water cylinder. This needs only to be fitted with an immersion heater and connected to the rising main and distribution pipes to provide a complete hot and cold water service.

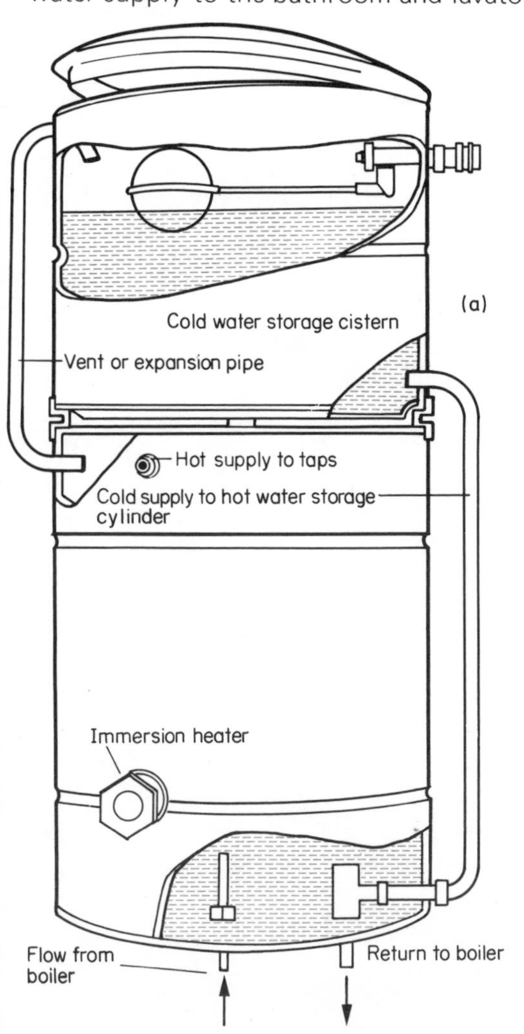

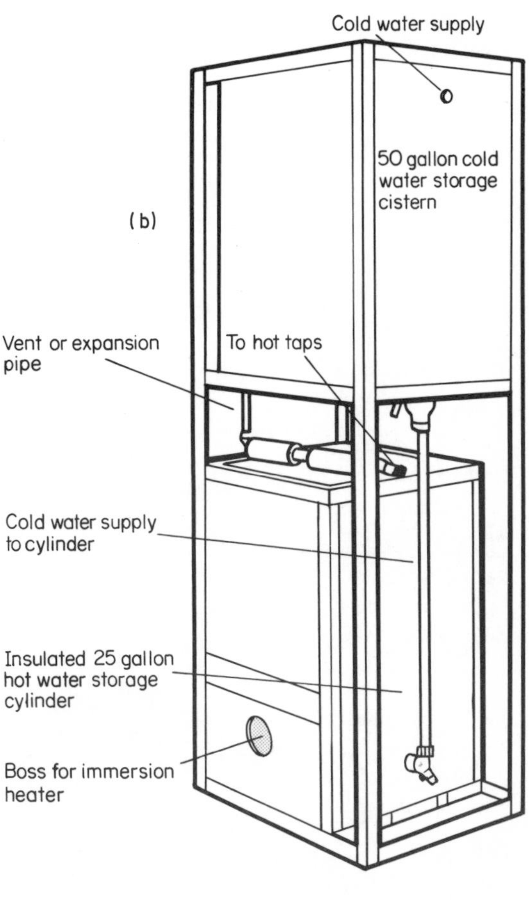

Later more sophisticated units were produced with a standard 50 gal cold water storage cistern. These could be placed in position, in a bathroom cupboard or airing cupboard, to provide a complete plumbing system, needing only the means of heating and the connection of the rising main and the hot and cold water distribution pipes.

Essentially these packaged or 'two-in-one' tanks are simple cylinder storage systems in which the cold water storage cistern and the hot water storage cylinder are brought into close proximity to form one unit. Pipe runs are accordingly shortened and, since the cold water cistern is immediately above the hot water cylinder, the risk of frost damage to the cistern is virtually eliminated.

The only disadvantage of the larger units of this kind is the fact that they normally have to be fitted at too low a level to provide sufficient pressure for a conventional shower.

Packaged or two-in-one systems—and some conventional systems—may dispense with a boiler of any kind and depend solely upon an electric immersion heater. In such cases the flow and return tappings will be blanked off. Provision must still be made for draining the cylinder when required. This is usually done by fitting a drain-cock on the cold water supply pipe, just before it enters the cylinder.

Faults in cylinder storage hot water systems

1. *Scale formation resulting from hard water*

When water containing dissolved bicarbonates of calcium or magnesium is heated to temperatures above about 140° F (60° C) carbon dioxide is driven off and the ***bicarbonates*** are changed into insoluble ***carbonates*** which form scale on internal boiler surfaces and on the elements of immersion heaters.

This results in delay in obtaining hot water and, with a boiler system, gurgling, hissing and knocking sounds from the boiler as overheated water forces its way through ever narrowing channels.

The scale insulates the metal of the boiler, and of the electric immersion heater, from the cooling effect of circulating water. Immersion heaters burn out and fail and, eventually, the metal of the boiler will burn through and a leak will develop.

Hot water systems can be descaled chemically by means of proprietory solutions introduced via the cold feed from the main cold water storage cistern. It is far better though, to prevent scale formation.

There are several ways in which this can be done. In hard water areas the immersion heater thermostat should be set at 140° F (60° C) and, where practicable, the boiler temperature maintained at this level. Water softening (see 'Hard Water Problems') or the introduction of chemical scale inhibitors into the cold water storage cistern are other means of controlling scale. It should be noted that chemical scale inhibitors, such as 'Micromet', do not ***soften*** water. Their action is to stabilise the chemicals causing hardness so that they do not precipitate out when heated.

Another method is to provide an ***indirect*** hot water system. An indirect cylinder storage system is illustrated. As can be seen an indirect system has a primary circulation, passing through the boiler, quite separate from the domestic hot water supply.

The primary circuit has a separate water supply from a small feed and expansion, or header, tank. Water in the storage cylinder is heated indirectly by a closed coil or heat exchanger from the primary circuit that passes through it.

In the primary circuit the same water is used over and over again, only the very small losses from evaporation being made up from the feed and expansion tank. Thus,

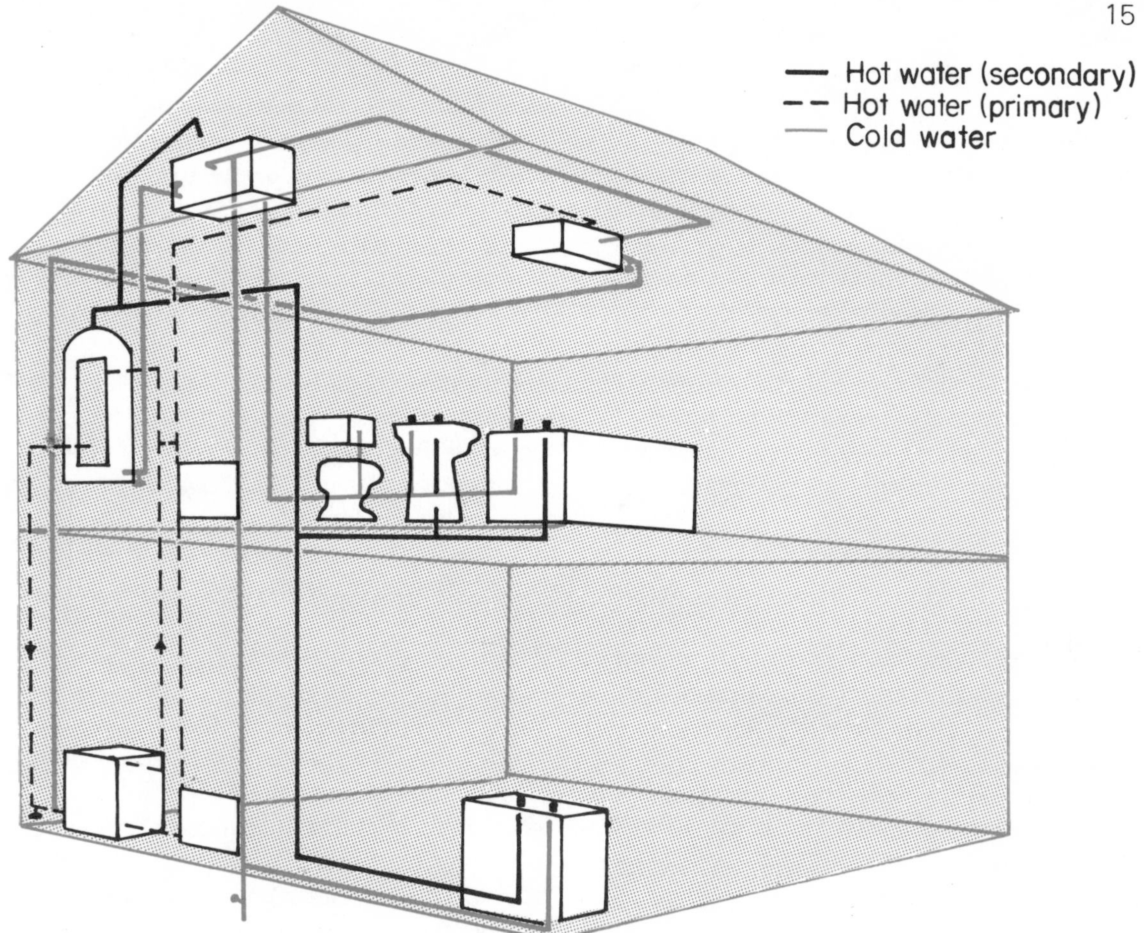

In an indirect hot water system, hot water for domestic use is heated by a closed heat exchange or calorifier within the specially constructed indirect cylinder. The water in the 'primary circuit' – circulating between boiler and cylinder – is used repeatedly, only the very small losses resulting from evaporation being made up from the small feed and expansion tank.
Water within the indirect cylinder is heated for domestic use by heated water circulating through a coil connected to the primary circuit.

when it is first heated, a small amount of scale is precipitated onto boiler surfaces. After that no more scale formation will occur.

Scale formation in the domestic hot water circuit will be minimal because the water in the outer part of the cylinder will rarely reach the high temperature at which scale formation takes place.

An indirect system also offers relative freedom from internal corrosion since the dissolved air, on which corrosion depends, is driven off when the primary circuit is first heated. Small amounts of air will however continue to be taken into solution via the surface of the water in the feed and expansion tank. It is wise therefore to introduce a chemical corrosion inhibitor into this tank.

It should be noted that with a conventional indirect system of the kind illustrated the primary circuit must be supplied from its own feed and expansion tank—never from the main cold water storage cistern. If the primary circuit of such a system is fed from the main storage cistern, mixing of the primary and domestic water will occur whenever the water in the primary circuit expands and contracts on heating and cooling. Fresh, hard and corrosive water will be drawn into the primary circuit and the advantages of an indirect system will be destroyed.

There are, on the market, patent 'self-priming' indirect cylinders that need no feed and expansion tank.

These appliances have a specially designed inner cylinder which, when the system is first filled, permits water to spill over from the domestic hot water into the primary circuit to fill it. A large air bubble, or air lock, then forms to prevent the return of the primary water. Provision is also made for the accommodation of the expansion of the water in the primary circuit when heated.

Some doubt has been expressed about the effectiveness of these systems in separating the primary from the domestic hot water. My own experience suggests that they are effective enough for general purposes provided that the water in the primary circuit is never allowed to boil and that there is sufficient space within the inner cylinder to accomodate the expansion of the particular primary circuit when heated.

An indirect cylinder should ***always*** be provided where hot water supply is to be installed in conjunction with even the smallest central heating system. Even where hot water only is required, an indirect system is recommended in areas where the water supply is hard or corrosive.

2. ***Rusty red water running from the hot water taps—particularly noticeable from the bath hot tap when a considerable volume of water has been drawn off.***

Check that the rust does not originate from the main cold water storage cistern. If it does, take appropriate action as suggested in Chapter 1.

If the cistern is free of rust the chances are that the rusty water results from corrosion within the boiler. The provision of an indirect system (see 1 above) is the only permanent answer but the use of a scale inhibitor such as Micromet in the cold water storage cistern may help.

3. *Air Locks*

The indications of an air-lock (poor and erratic flow from a hot tap, often accompanied by hissing and spluttering) are sometimes mistakenly attributed to scale formation in taps or supply pipes. Air locks result from bubbles of air, trapped in the supply pipe, preventing or restricting the free flow of water.

When a primatic system is filled for the first time water spills over from the outer cylinder into the primary circuit (1) but is prevented from returning by an air-lock that forms in the inner cylinder (2). On being heated, water in the primary circuit expands, pushing the air in the air-lock down into the lower hemisphere of the inner cylinder (3).

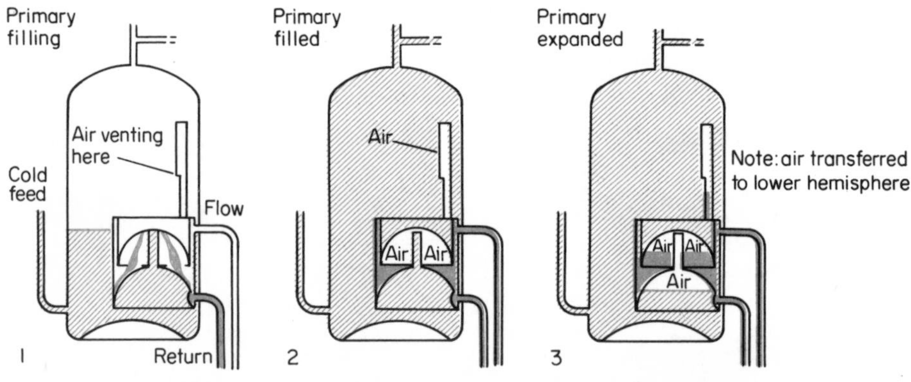

They can usually be cleared by connecting one end of a length of garden hose to the cold tap over the kitchen sink (this is supplied direct from the main) and the other end to the tap giving trouble. Turn both taps on full and the mains pressure should blow the air bubble out of the system.

Always seek the cause of recurring air locks. Perhaps the most common cause is having a pipe of too small a diameter taking cold water from the storage cistern to the hot water cylinder.

If this pipe is only 15 mm (½ in) in diameter it will be incapable of replacing water drawn off from the ¾ in bath hot tap. As a result the level of water will fall in the vent pipe and eventually, bubbles of air from the vent pipe will be drawn into the hot water supply pipe.

If the supply pipe *is* of the right diameter check that any control valve fitted into it is of the same size. A 15 mm control valve fitted into a 22 mm (¾ in) supply pipe will effectively reduce the diameter of the pipe to 15 mm. Check too that any such valve is fully open.

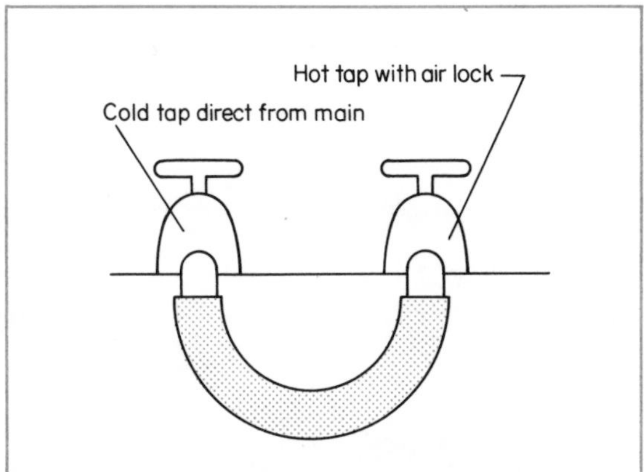

It is usually possible to clear an air-lock by connecting the tap giving trouble to the cold tap over the kitchen sink. When both taps are turned on, the water from the sink tap being under mains pressure forces the air bubble out of the system.

Other possible—though less probable—causes of air-lock are too small a cold water storage cistern or a sluggish ball-valve supplying this cistern.

'Horizontal' runs of pipe connected to the main vent pipe should always fall slightly away from the vent so that any air bubbles can escape.

4. ***Loud bubbling noises, possibly accompanied by the sound of water pouring out of the vent pipe into the cold water storage cistern,***

This again, is the result of a form of air-lock. The flow pipe from the boiler should ***rise*** all the way to the cylinder flow tapping. The vent pipe from the dome of the cylinder should ***rise*** throughout its length.

If, in either of these pipes, there is a horizontal run or, worse still, a slight

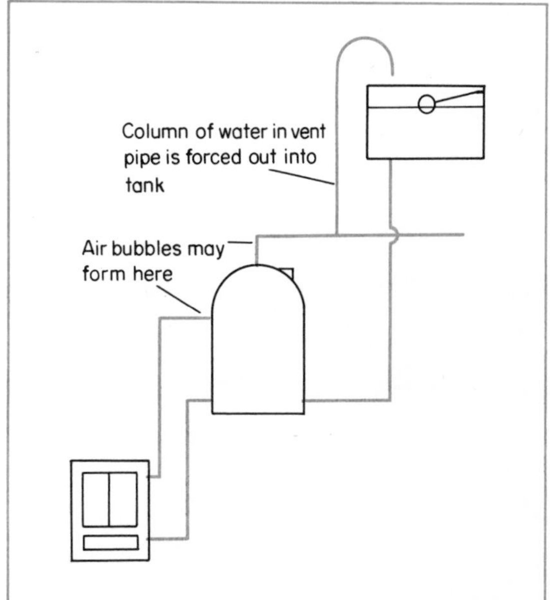

If the pipes indicated do not slope slightly upwards air bubbles will collect in the horizontal lengths. Pressure will eventually force these air bubbles out of these pipes and they will escape via the vent pipe, noisily pushing the column of water in the vent pipe in front of them.

back fall, dissolved air—driven off in bubbles from the boiler—will collect at this point.

Pressure will build up behind this bubble until it is sufficient to drive it out of the pipe. It will then, if sufficiently large, push the water standing in the vent pipe out into the storage cistern.

5. *Reversed Circulation*

This may occur during the summer when the boiler is not in use and hot water is being supplied by immersion heater only. A wasteful—and very expensive—circulation may take place ***down*** the flow pipe to the boiler and back, via the return pipe, to the cylinder.

A probable cause is having the cylinder more or less on the same level as the boiler. The best position for the storage cylinder is close to the boiler but at a higher level.

If the cylinder cannot be raised the situation can be remedied by realigning the flow pipe so that it rises ***inside*** the cylinder insulation to the flow tapping.

An even more serious form of reversed circulation will occur when a door is interposed between boiler and cylinder and the flow pipe is taken, as indicated, over the door. When the boiler is out of use electrically heated water will inevitably rise up the vent pipe and, cooling, descend to the cold boiler.

Moving the position of the cylinder is the only really satisfactory solution to this problem. A second-best remedy is to extend the flow pipe so that it enters the cylinder below the level of the immersion heater. This involves the provision of an additional vent pipe. It will effectively prevent reversed circulation but will result in delay in the cylinder heating up when the boiler is in use.

(a) Reversed circulation can result in high electricity bills. The remedy in this case is to take the flow pipe to the cylinder within the cylinder insulation.

(b) Reversed circulation resulting from this layout is less easily cured. The best solution is to redesign the system so that there is no door interposed between boiler and cylinder but a 'second best' remedy in indicated.

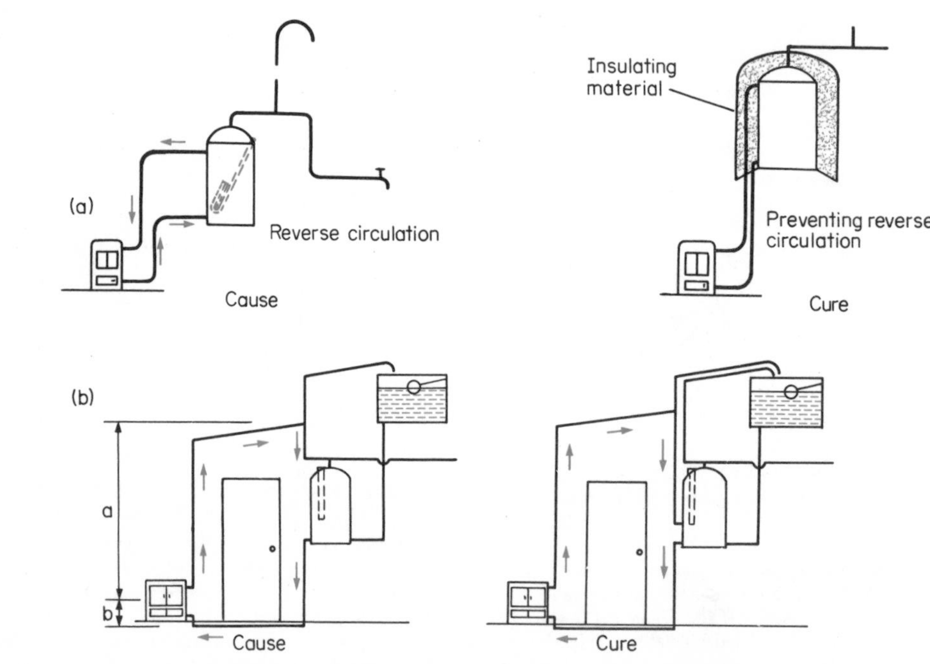

Chapter 3 Water heating by electricity and gas

In the previous chapter I mentioned that a cylinder storage hot water supply might be provided with an electric immersion heater, either to supplement a boiler using some other fuel, or as the sole source of hot water supply.

Water heating by electricity

Most manufacturers of electric appliances manufacture cylinders, complete with immersion heaters, designed as a sole source of domestic hot water. These are usually intended for installation under the draining board of the kitchen sink and, for this reason, are often called 'under draining board' or UDB heaters.

A feature of appliances of this kind is very heavy, built-in, insulation. They are most usually provided with two, horizontally aligned, electric immersion elements. These take advantage of the fact that hot water 'floats' on top of cold.

The upper element is kept permanently switched on to provide the relatively small amounts of water required for washing up, washing and shaving and so on. It heats only the water in the cylinder above the upper

(a) The upper immersion heater of this unit is switched on continuously to ensure that there is always sufficient hot water for personal washing, washing up etc. The lower heater is switched on an hour or so before a larger volume of water is required for, for instance, baths or domestic laundry.

(b) This off-peak water heater switches on at night, when electricity charges are lower, and switches off during the day. It has been estimated that 50 gal of hot water is sufficient for the daily needs of an average family. The tall, slim design ensures stratification of the heated water.

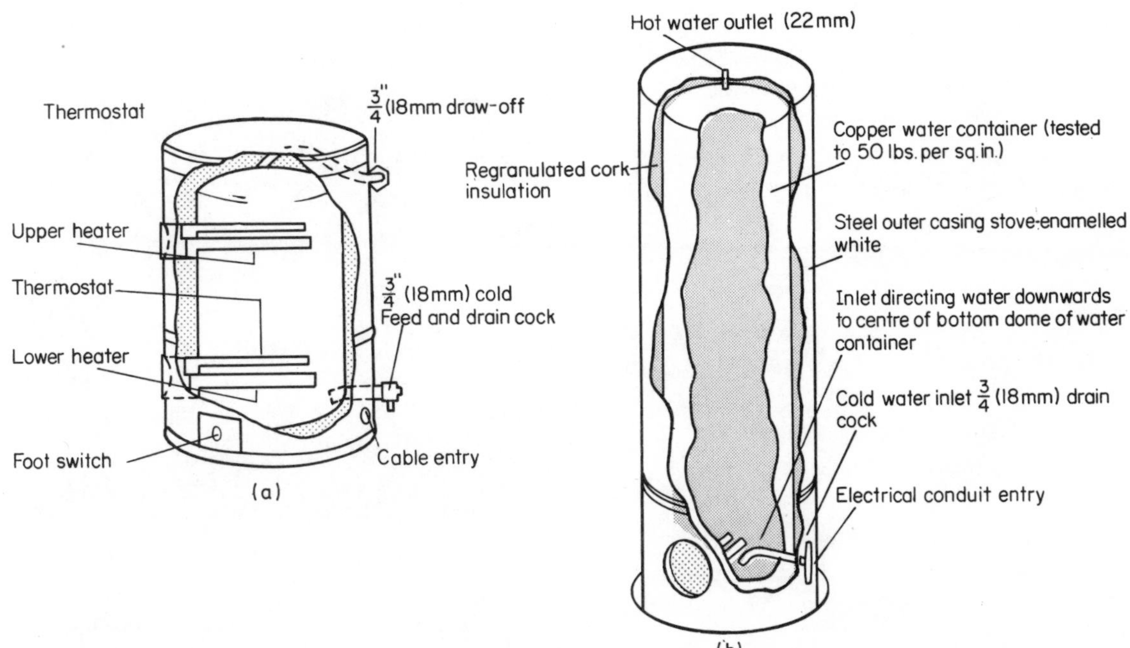

element. The lower element is intended to be switched on an hour or so before greater volumes of hot water are required for baths or laundry purposes.

A variation on the same theme is the off-peak electric water heater designed to take advantage of the cheaper off-peak electricity charges. Typically a heater of this kind is tall and slim to encourage the stratification of the heated water. A special 'spreading' device is provided for the cold water inlet at the base to ensure that incoming cold water spreads evenly over the lower part of the cylinder, pushing the heated water upwards without mixing with it. These off-peak cylinders are also provided with very heavy built-in insulation.

They customarily have a capacity of 50 gal--twice that of the average UDB heater. It has been estimated that 50 gal of hot water meets the daily needs of a normal family. The heater is switched on overnight, to take advantage of the off-peak rates. During the day the electric element is switched off and the stored, heated water used by the family.

Essentially UDB and off-peak heaters of this kind are cylinder storage systems specially adapted to make the best use of electricity as a means of heating. Most of them need a separate cold water storage cistern but there are 'two-in-one' versions—usually called 'cistern heaters'—which incorporate their own small, cold water supply cistern in the upper part of the unit.

Such cistern heaters must be situated above the level of the highest hot water draw-off point. Their advantages and disadvantages are identical to those of the smaller two-in-one or packaged hot water systems described in Chapter 2.

Open-outlet electric water heaters, frequently installed over sinks and wash basins, operate on a rather different principle. These are designed for connection direct to the main and an essential part of their

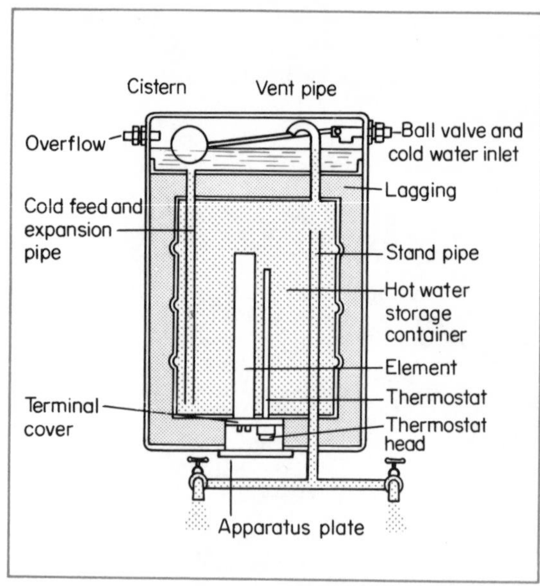

This is essentially a simple, cylinder hot water system especially designed for use with electricity as the sole source. The small cistern is, of course, not large enough to supply any cold water fitting. These must be supplied from another cistern or from the main.

The open-outlet electric water heater has the control valve on the inlet side of the heater. Cold water flowing into the appliance displaces heated water which overflows down a stand-pipe to the outlet spout. (courtesy of Heatrae Ltd)

design is the position of the control valve or tap. This must be on the inlet side—*not* on the outlet—of the appliance.

These units have a vertically aligned immersion heater inserted through the base. When hot water is required the inlet control is opened. Cold water flows in at the base and the hot water within the unit overflows through an internal stand-pipe connected to the outlet spout.

Modern variations of this appliance may be installed under, instead of over, the sink or wash basin. These must still comply with the essential requirement of a controlled inlet and a free outlet.

In recent years a number of electric 'instantaneous' water heaters have come on the market. These too are connected directly to the water main and have their control valve on the inlet side of the appliance. Used mainly for spray hand washing and for the provision of showers in situations where this would otherwise be impossible, they heat water 'instantaneously' as it passes through electrically heated channels within the appliance.

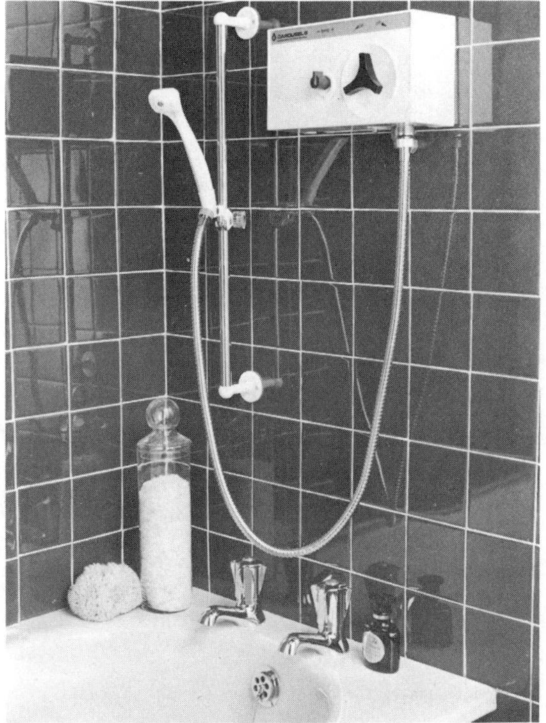

Instantaneous electric water heaters can be useful for the provision of a supply of hot water to a shower of a wash basin where a more conventional means of supply would be difficult or uneconomic. They can give trouble from scale formation in hard water areas.

They have the advantage that electricity is used only to heat water actually drawn off and are therefore particularly economical in situations where use of hot water is occasional only.

Their disadvantages include a rather low rate of delivery, scale trouble in hard water areas and the fact that they have not yet received the universal approval of Water Authorities and Electricity Boards. The need for energy conservation may well lead to further development of this kind of appliance.

Water heating by gas

In the field of instantaneous water heating gas has, so far, all the advantages. Instantaneous gas water heaters are available for installation over baths, sinks and wash basins and larger 'multipoint' models can provide a whole-house domestic hot water supply. Normally supplied direct from the main they make it possible, where this is permitted, for the cold water storage cistern to be dispensed with entirely.

The problem of the disposal of the flue gases, which invested the Edwardian 'bathroom geyser' with its potentially lethal qualities, has been solved by the invention of the 'balanced flue'.

The combustion chamber of balanced flue gas appliances is completely sealed off from the room in which the appliance is situated. Air to support combustion is drawn from an inlet through the external wall behind the appliance. The flue outlet is immediately adjacent to this inlet.

Thus, the system is balanced. If a gale force wind is blowing against the wall in which the flue outlet is situated, it will

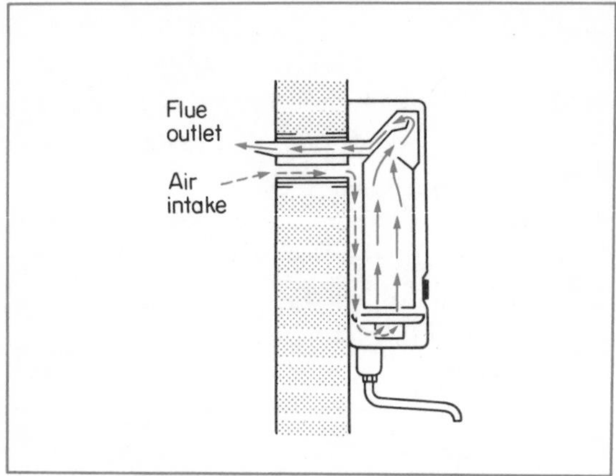

Balanced flue gas appliances are available for space or water heating and can be installed against any external wall. The combustion chamber is sealed off from the room in which the appliance is fitted. Air intake and flue outlet are adjacent and are therefore 'balanced'.

be blowing equally against the fresh air inlet. Normal combustion will be unaffected.

Instantaneous heaters are not, of course, the only means by which gas can be used to provide domestic hot water supply. There are gas over-sink storage heaters similar to electric free outlet heaters. Large gas-fired boilers, used in conjunction with indirect hot water systems (see previous chapter), can provide both hot water supply and central heating. Small gas boilers, or 'circulators', usually fitted closely to the walls of a storage cylinder, operate in the same way as a solid fuel boiler to provide hot water supply only.

Faults in electric and gas hot water systems

These systems are, on the whole, extremely efficient and trouble free. Complaints are more likely to relate to excessive fuel bills than to inefficient operation. If your bills are consistently higher than those of friends or neighbours with similar systems, consider the following points. Any one of them could be the cause.

1. Inadequate insulation of storage cylinder
An uninsulated copper cylinder 18 in in diameter and 36 in high (nominal capacity 30 gal) will lose 86 units of electricity per week if the air temperature is 60°F (15°C) and the temperature of the water in the cylinder is maintained at 140°F (60°C). If the water temperature is maintained at 160°F (71°C) it will lose 115 units per week.

Work out the money wasted each week at the current price of electricity per unit!

Efficient lagging is the answer. Thickness of lagging material is more important than the nature of the material used. Optimum thickness is 3 in. The cylinder above will lose only six units of electricity per week if maintained at 140°F (60°C) and lagged with a 3 in thickness of glass fibre. If the thickness is reduced to 2 in, 8.8 units will be lost each week.

Never strip off part of the lagging to heat an airing cupboard. It is cheaper to install a small low-output electric heater in the cupboard.

2. Long 'dead legs'
In most modern homes the bathroom and kitchen are in close proximity. In a two storey house the bathroom will probably be above the kitchen. In a bungalow it is likely to adjoin the kitchen.

This keeps pipe runs—'dead legs'—from the cylinder to the draw-off points, short. Long dead legs waste heat. When the tap is turned off the dead leg is full of water that you have paid to heat. It will cool rapidly.

A dead leg of 15 mm (½ in) copper tubing carrying water at 140°F (60°C) to a sink or basin tap will waste about 0.19 units per foot run per week. A similar 22 mm (¾ in) copper tube will waste 0.38 units per foot run per week.

Where the hot water storage cylinder is in, or adjacent to, the bathroom and the kitchen sink is more than, say 20 ft away, it is worth considering providing a separate,

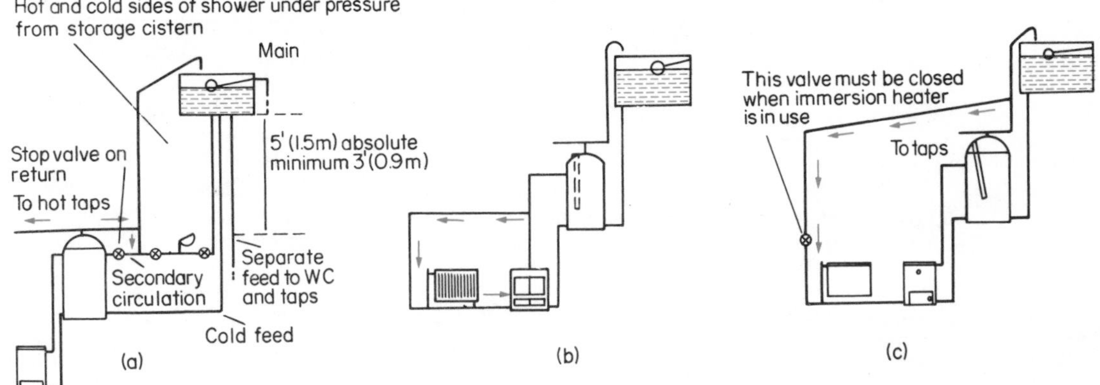

(a) A secondary circulation is sometimes provided on the hot water supply to — for instance — a shower, to reduce the delay in the arrival of the hot water. This secondary circulation must be stopped when the water is heated by electricity.

(b) Circulation of electrically-heated water through a towel rail or radiator will result in high electricity bills. Where such a circulation is provided it should either be wholly below the level of the immersion heater or, alternatively, a stop-valve (c) must be provided so that circulation can be stopped when the immersion heater is in use.

small water heater, over the kitchen sink.

3. Circulating piping

Electrically heated water should never be permitted to circulate. Sometimes, for instance, in order to speed up delivery of hot water to a shower some distance from the cylinder, a secondary circulation, as illustrated, is provided.

If you have a secondary circulation of this kind see that it is fitted with a stop-valve on the return run and that this valve is turned off when the water is being heated electrically.

In Chapter 2 we saw that even the smallest central heating system required the provision of an indirect cylinder. Yet it is quite permissible to run a single heated towel rail from a direct hot water system. This is, in fact, common practice.

Electrically heated water must never be permitted to circulate through such a towel rail.

Where practicable the towel rail circulation should be taken from the flow pipe from boiler to cylinder below the level of the immersion heater. Where this is impracticable fit a stop-valve into the towel rail circuit and see that it is turned fully off when the immersion heater is switched on.

Electrically heated water circulating through a 15 mm (½ in) copper tube at 140°F (60°C) will waste 1.36 units of electricity ***per foot run*** per week. If, as is possible with a towel rail circuit, the tubing is 28 mm (1 in) in diameter the wastage will be 2.33 units per foot run per week. (See also 'Reversed Circulation', Chapter 2).

4. One-pipe circulation

If the vent pipe from the cylinder rises vertically upwards throughout its length to the cold water storage cistern, there will be appreciable heat wastage from one-pipe circulation.

Currents of heated water will rise, by convection, up the centre of the 22 mm (¾ in) vent pipe. Cooling, they will descend against the inner walls of this tube.

One-pipe circulation can be prevented by taking the vent pipe ***almost*** horizontally from the dome of the cylinder for a distance of about 18 in before permitting it to rise vertically upwards to the cold water storage cistern.

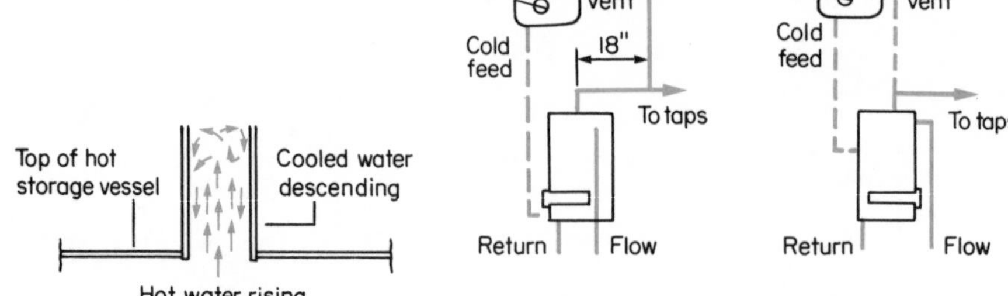

Heat wastage from 'single pipe' circulation will result if the vent pipe rises vertically from the dome of the hot water cylinder. This can be prevented by taking the vent pipe horizontally for 450 mm (18 in) before it rises to the storage cistern.

Other faults in electric and gas hot water systems

1. Faults common to all cylinder storage hot water systems

Cylinder storage hot water systems operated by gas or electricity are, of course, prone to air-locks and other faults affecting all cylinder storage systems (see Chapter 2).

2. Poor recovery period after hot water is drawn off from storage cylinder

Check that thermostat of electric immersion heater is correctly set—140°F (60°) in hard water area, 160°F (71°C) in soft water area.

Trouble could be due to build-up of scale on electric immersion heater. Electrically heated cylinder systems can be descaled chemically in the same way as simple cylinder systems operated by a boiler (See previous chapter). Aim at scale prevention —correct thermostat setting, use of scale inhibitor in cold water storage cistern, water softening (see Chapters 2 and 9).

3. Poor flow or inadequately heated water from an instantaneous gas water heater

This trouble could again be caused by scale formation in the water channels.

Descaling a large multi-point instantaneous heater is hardly a d.i.y. job but smaller appliances can usually be descaled as follows.

Cut off the water and gas supply. Disconnect the water inlet to the heater. Connect a length of rubber hose or other rubber tubing to the water inlet and insert a glass funnel in the other end of the tube. Raise funnel to above the level of the top of the heater and fix in position. Pour in descaling fluid. ***Pour slowly and carefully,*** looking out for foaming back up the tube.

When faced with a poor flow or inadequately heated water from an instantaneous water heater don't overlook the natural limitations of these appliances. They cannot heat a given volume of water to a predetermined temperature. What they ***can*** do is to raise a given volume of water ***through a predetermined range of temperatures***. In prolonged cold weather either outlet temperature of the water, or flow, will be reduced.

5. Drip from spout of free outlet water heater, particularly when heating up.

This trouble is usually assumed by the householder to be due to a faulty washer in the inlet control valve. It is however more likely to be caused by scale formation in the siphon device at the top of the standpipe outlet inside the appliance.

The purpose of this siphon is to lower the water level in the appliance to a level ½ in or so below the rim of the stand-pipe when the control valve is turned off.

This ½ in is to accommodate the expansion of the water in the appliance as it heats up. If the siphon becomes clogged with scale it will fail to operate. The appliance will fill to the rim of the stand-pipe when cold. There will then be a constant drip, resulting from expansion, as the water heats up.

Once again, descaling is the remedy. Small open outlet appliances can be descaled as described in 3 above for small instantaneous heaters.

Chapter 4
Taps, stop-cocks and ball-valves

Taps, stop-cocks and ball valves—or float-valves as they are increasingly called—have in common the fact that they are all fittings designed to control the flow of water through pipes.

Taps

Looking through an illustrated catalogue of modern taps it is easy to forget that, despite the 'tomorrow's world' appearance of most of the taps displayed, they all—with one exception—operate on exactly the same principle as the old crutch-headed bib-cocks familiar half a century ago. In fact, of course, even today straightforward 'crutch' or 'capstan' headed bib and pillar taps are manufactured and sold. The great majority of taps in British homes are still of this traditional pattern.

Bib-taps have a horizontal inlet. Twenty years ago this is the kind of tap that would have been found in any suburban home, projecting from the tiled wall above the ceramic kitchen sink. The advent of the enamelled steel and stainless steel sink unit has relegated the more basic kinds of bib-tap to the outside wall, to provide a garden and garage water supply. Variations on the bib-tap theme are however still to be found over some wash basins and baths, particularly where space is limited or concealed plumbing it particularly required.

Pillar taps, with a vertical inlet, are nowadays fitted almost universally in the holes provided for them in modern sinks, wash basins and baths.

Both kinds of tap work in the same way. Turning the tap-handle clockwise forces the valve—or jumper—complete with its composition washer, down on to the valve seating to close the waterway. Turning the handle anti-clockwise frees the valve and allows water to pass out of the tap.

Bath, sink and basin mixers are simply two taps with a common spout. Sink mixers have a rather different design from bath and basin mixers because it is both illegal and impracticable to mix in any plumbing fitting, water from the main (the cold supply to the kitchen sink) and water from a storage cistern (the hot supply). This difficulty is overcome by providing separate channels for the hot and cold water within the spout of the mixer. Hot and cold streams of water mix *in the air* as they leave the nozzle.

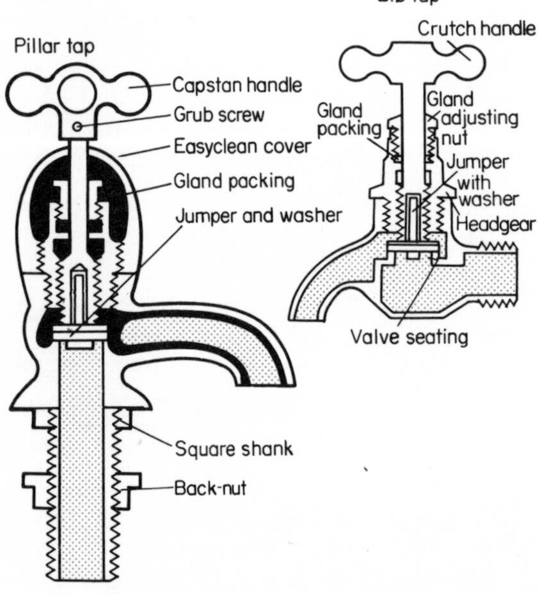

Bib (horizontal inlet) and pillar (vertical inlet) taps of this kind provide the basic design for practically all domestic taps, despite the vastly different appearance of some modern models.

The one exception in design to which I referred is Bourner's 'supatap'. The supatap, renowned for its rapid washer change without the need to cut off the water supply, gives a particularly smooth flow of water and has an appearance that many find attractive.

The tap is opened and closed by rotating its nozzle, provided with 'ears' for that purpose. A failing of early supataps was the fact that the metal ears of a hot tap became uncomfortably hot as the tap was used. Later models have ears of kemetal plastic which has excellent insulating qualities.

To connect a hose to a supatap a special hose adaptor is essential. This attaches to the ears of the tap to permit the nozzle of the adaptor to remain stationary while the body rotates with the tap nozzle.

Taps have not, at the time of writing, been subjected to the benefits, or otherwise, of metrication. They are still catalogued as ½ in or ¾ in. Bath taps are normally ¾ in and the ½ in size is used for sinks, basins and garden water supplies. ½ in taps are, of course, designed to connect to modern 15 mm copper tubing and ¾ in taps to the 22 mm size. It is probable that, when taps are metricated, these are the nominal metric sizes by which they will be designated.

How to fit a tap

To fit a pillar tap into a ceramic sink or basin, unscrew the back-nut and slip a plastic washer over the threaded 'tail'. Insert in the hole provided. Slip another plastic washer over the tail and tighten up the back nut. This nut should be tightened sufficiently to hold the tap securely but don't overtighten. Ceramic surfaces are easily damaged.

When fitting a pillar tap into a sink, bath or basin of stainless or enamelled steel or other thin material, the same procedure is adopted except that a special spacer or 'top hat' washer must be provided between the back-nut and the under surface of the fitting to accommodate the protruding shank of the tap.

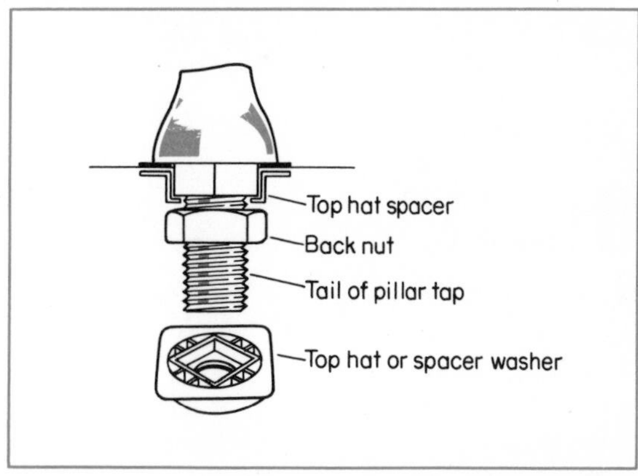

When fitting a tap into a thin material – for example, the stainless steel surface of a sink or vanity unit – a 'top hat' or spacer washer must be provided to accommodate the protruding shank of the tap.

Faults in taps

1. Dripping after the tap has been turned off

This is an indication that the washer needs to be renewed. Set about it this way:

With any tap other than a supatap, cut off the water supply. For the cold tap over the kitchen sink turn off the main stop-cock (see Chapter 1). Other taps *may* have stop-valves in the pipe-lines supplying them with water but the probability is that they will not.

To cut off the water supply to these taps you will therefore have to tie up the float arm of the ball-valve supplying the main cold water storage tank. This will prevent water flowing into this tank and permit you to drain it from the bathroom taps.

Provided that the bathroom cold taps are supplied from the main storage tank and not from the main you need not drain off the hot water stored in the cylinder, even if it is the washer of a hot tap that needs renewal. Open up the bathroom cold taps and leave open until water ceases to flow. Then open up the hot tap that needs to be rewashered. A pint or so of water will flow out of it and flow will then cease. This is because the supply to the hot taps is taken from ***above*** the storage cylinder.

Unscrew the easyclean cover of the tap. You should be able to do this by hand. If you are compelled to use a wrench, pad its jaws to protect the chromium plating.

Raise the easyclean cover and you will see the large hexagon nut with which the top-gear of the tap is screwed into the body. Hold the body of the tap firmly with one hand, or with a padded wrench, grasp the hexagon nut with another wrench and unscrew so as to remove the top-gear.

If you are dealing with the cold tap over the kitchen sink you will probably find the jumper with washer attached resting freely on the valve seating inside the tap body. Other taps will probably have the jumper pegged into the head-gear, so that it can be rotated but cannot be withdrawn.

To renew the washer you unscrew the small retaining nut, fit the washer and replace the nut.

This may be easier said than done. It is quite on the cards that the nut will have become so scaled up and corroded as to be all but unmoveable. With a free, un-pegged, jumper there is an easy answer. Fit a new jumper complete with washer. You can get these at any household or d.i.y. store.

If the jumper ***is*** pegged into the head-gear you have a more difficult problem. Make a real effort to unscrew that nut. apply a drop of penetrating oil and try again after fifteen minutes or so.

If the nut really cannot be moved, force the jumper out of the head-gear with the blade of a screw-driver, breaking the pegging. Fit a new jumper and washer

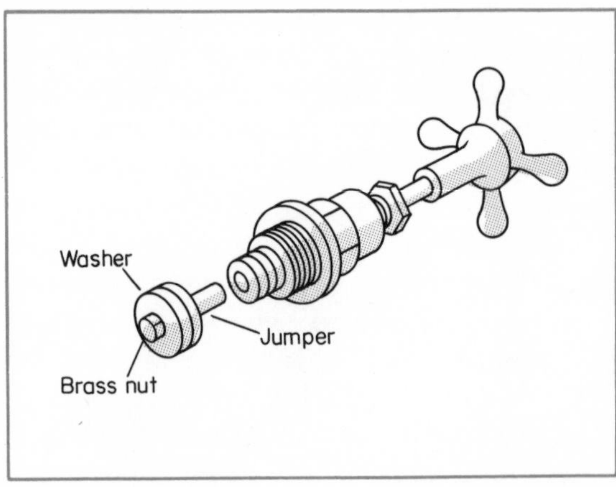

Some taps have the jumper 'pegged' into the headgear of the tap. The jumper will turn round and round but cannot readily be removed.

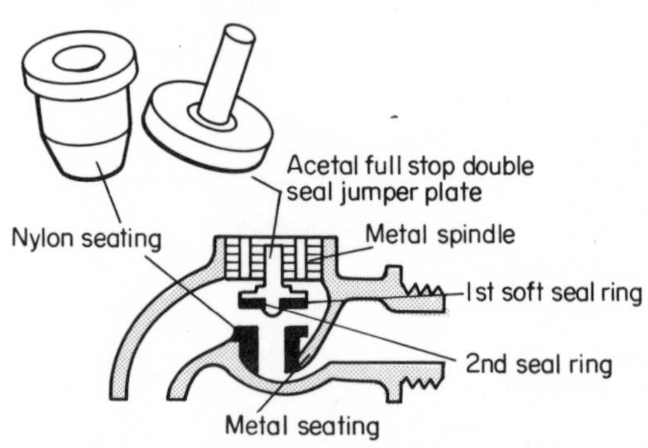

The 'Full Stop' tap washer and seating set provides a simple d-i-y method of 'resting' a tap. The nylon seating is placed in position on the existing valve seating and the head-gear is then screwed down hard.

complete but, before you slip it into position, burr the stem of the jumper so that it fits fairly tightly into the headgear.

Continued dripping after rewashering indicates that the valve seating has become scored and damaged. It is no longer providing a watertight joint. The easiest solution to this problem is to fit one of the new plastic valve seating and jumper sets. These are forced down onto the existing valve seating to give a watertight connection.

My own experience suggests that a tap fitted with one of these plastic valve seatings may continue to drip for a time. This stops as the tap is used and screwed down hard so as to force the new seating tightly against the old one.

How about modern taps with shrouded heads? With these the handle and head-gear appear to comprise a single unit.

To dismantle these taps prize off the small plastic 'hot' or 'cold' indicator in the centre of the shrouded head. Under this you will find the head of a screw that keeps the head in position. Undo this screw and the head can be removed, revealing the interior of the tap to be little different from that of a conventional pillar or bib tap.

Nothing could be simpler than rewashering a Supatap. Don't turn off the water supply. Unscrew and release the retaining nut directly above the nozzle. Then turn the tap on—and on. At first the flow of water will increase but it will cease as the check-valve falls into position. Continue to unscrew until the nozzle comes off in your hand.

Tap the nozzle on a hard surface to loosen the anti-splash device. Then turn the nozzle upside down and this anti-splash, which holds the washer and jumper, will fall out.

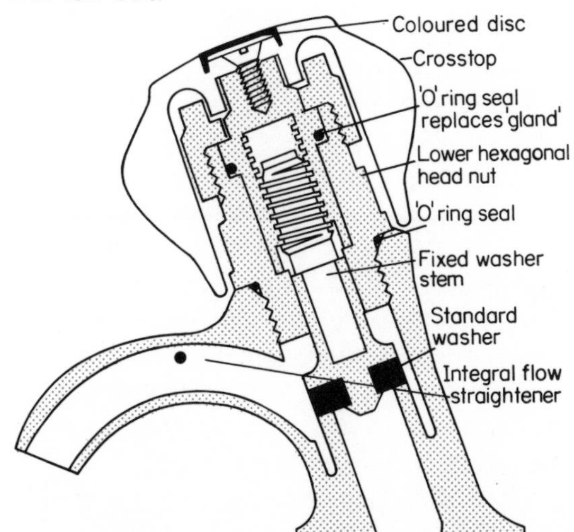

Opella 500 series pillar taps are suitable for standard baths, basins and sinks. They are of kemetal plastic construction and have an O-ring ring seal instead of the conventional gland.

Renew the washer and jumper, replace in the nozzle and screw the nozzle back on to the body of the tap. As you do this, don't forget that the nozzle has a ***left-hand thread***. Turn anti-clockwise to replace it. Finally, reconnect the retaining nut to the top of the nozzle.

2. Water leaking past the spindle of the tap when the tap is turned on.

This is due to failure of the gland packing and is most likely to occur in older pattern taps with a traditional 'gland'. Connecting a hose to a kitchen tap may produce this trouble by creating back-pressure within the tap.

Another possible cause is detergent charged water dripping from the hands, running down the tap spindle, and washing the grease out of the gland packing. It was to prevent this—as much as for improved appearance—that the all-protecting 'shrouded head' was developed.

First of all try tightening the gland adjusting nut. This is the first nut through which the spindle of the tap passes. You will probably have to remove the crutch or capstan head and the easyclean cover to get at it.

To remove the head or handle unscrew and remove the tiny grub screw holding it in place. The head may then come off easily. If it does not, raise the easyclean cover as far as it will go and jam a piece of wood, or two clothes pegs, between the bottom of the cover and the body of the tap. Screw the tap down and the upwards pressure of the easyclean cover will force off the head.

Give the gland-adjusting nut half a turn or so in a clockwise direction. This may cure the trouble.

Eventually, of course, all the adjustment will be taken up and the gland will have to be repacked. To do this unscrew and remove the adjusting nut to reveal the gland

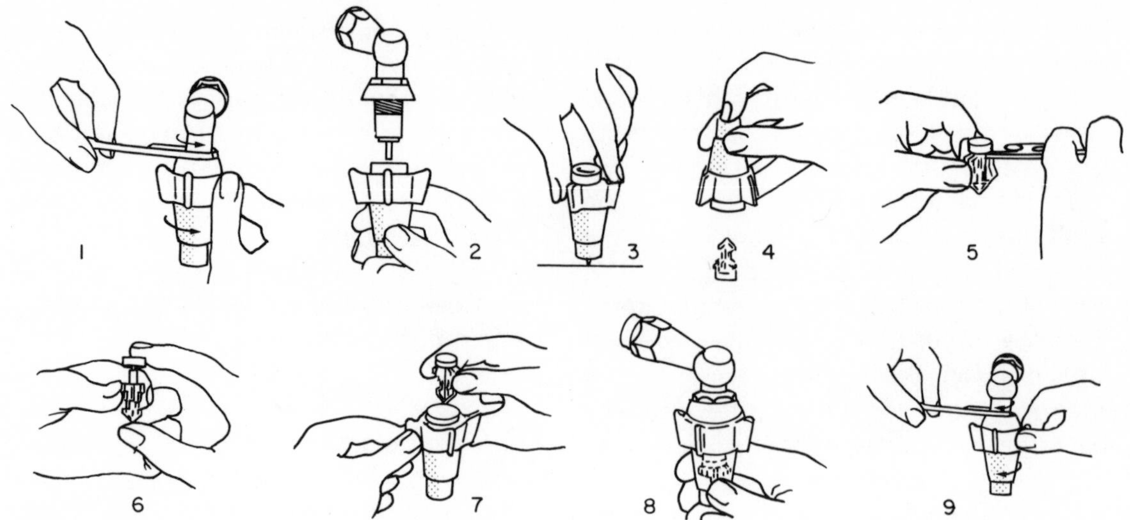

Supatap washer change

1. *Partly open nozzle and undo tap nut completely (in direction of arrows). Water will commence to flow.*
2. *Continue opening nozzle until it is detached. Flow will be reduced to a trickle before nozzle is completely free.*
3. *The centre of the anti-splash protrudes below the nozzle. Press this against a hard surface (not the basin) to release the anti-splash from nozzle.*
4. *Press out the anti-splash and jumper from the nozzle.*
5. *The jumper may be fast in the anti-splash but can be levered out with a coin or a blade.*
6. *Replace a new washer-jumper into the anti-splash and make sure that the pin clicks into the top hole of the anti-splash.*
7. *Put anti-splash, complete with washer jumper into the nozzle with the washer uppermost.*
8. *Ensure the washer jumper is in position indicated by dotted lines and screw nozzle back into shank.*
9. *As the nozzle is being screwed onto the shank the water supply will again commence. Continue screwing until almost closed then tighten top retaining nut and close tap completely, turning in direction of arrows.*

chamber. Pick out, with the point of a penknife, all existing packing material. Repack the gland with household wool steeped in petroleum jelly. Pack down tightly and screw the gland adjusting nut back into position.

3. Water hammer

Water hammer—heavy banging or vibration in the supply pipes—especially noticeable when a tap has been turned off, is due to shock waves resulting from the sudden stoppage of a flow of water.

Faulty gland packing is a common cause. As water escapes past the spindle the tap becomes easier and easier to turn. Eventually it may be possible to spin it on and off with a flick of the fingers.

This will inevitably produce water hammer. The remedy is as suggested in *2* above.

A faulty or unsuitable ball-valve (see 'ball-valves') is another common cause of water hammer.

Stop-cocks

Stop-cocks or stop-valves are essentially a form of screw-down bib tap, set in the run of water pipe, to stop or control the flow of water at will. They may be provided with compression joint inlets and outlets for connection to copper, stainless steel or polythene tubing, screwed inlets and outlets for connecting to steel tubing or plain ends for soldering to lead pipe.

When fitting a stop-cock it is essential to ensure that the arrow engraved on the body of the fitting points in the same direction as the flow of the water. If a stop-cock is fitted the wrong way round, water pressure will force the washer and jumper on to the valve seating and prevent the flow of water even when the valve is open.

There are now tiny and unobtrusive stop-cocks, operated by a screwdriver, that can be fitted into the water supply pipe-line immediately behind any tap or ball-valve. In this position they enable the water draw-off point to be isolated for washer changing or other servicing without the need to drain the system or to cut off the supply to any other plumbing fitting.

(a) A screw-down stop-cock resembles a bib-tap set into the run of a pipe. The arrow engraved on the body indicates the direction of water flow.

(b) The Markfram mini-stopcock is turned on and off by means of a screw driver. Fitted in a water supply pipe close to a tap or ball-valve it permits those fittings to be changed or serviced without interrupting water supply to other draw-off points.

Faults in stop-cocks

1. Washer failure and leakage past spindle

These troubles can occur in stop-cocks as well as in taps. Remedies are indicated above. Gland failure, indicated by leakage past the spindle, should receive immediate attention. Leakage on to a boarded floor, in the confined ill-ventilated spaces in which stop-cocks are frequently found will, almost inevitably, introduce dry rot into the home.

Where the water supply needs to be cut off to the main stop-cock it may be necessary to ask the Water Authority to turn off ***their*** stop-cock in the highway.

2. Stop-cock jammed through long disuse

This is a very common failing of stop-cocks. It is usually discovered only when a plumbing emergency occurs. The stop-cock is found to be immoveable.

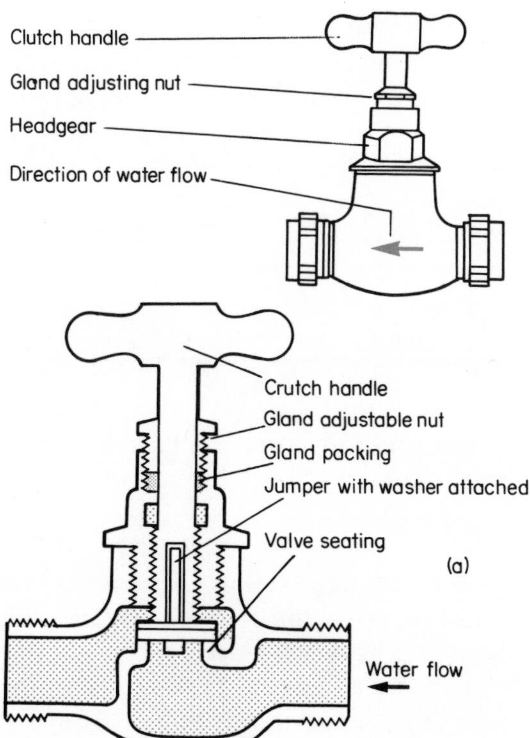

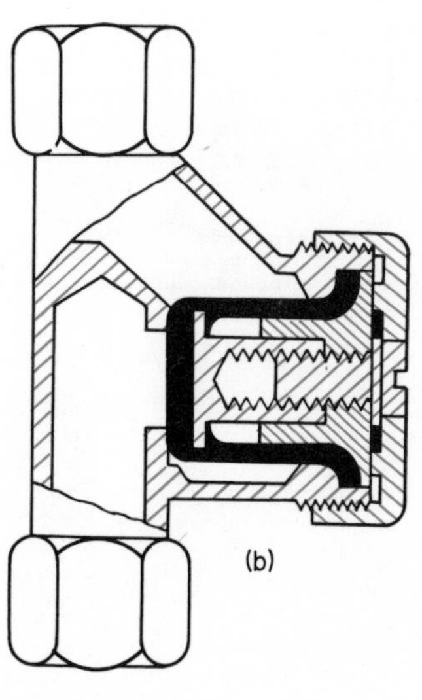

This is a trouble easier to prevent than to cure. It is a good idea to open and close all the household stop-cocks two or three times, at least twice a year. This will make sure that they are in working order when the need arises.

Ball-valves

A ball-valve, or float-valve, can be regarded as a tap with a float-operated control, designed to maintain water at a constant level in a water storage or lavatory flushing cistern.

There have been some interesting developments in ball-valve design in recent years and even in recent months. It seems at least possible that the diaphragm/equilibrium valve on the lines of the 'Torbeck' valve produced by Ideal Standard Ltd. during 1974, will set the pattern for the future.

Undoubtedly however, ball-valves of traditional pattern will be found in British homes for many years to come.

The simple, sturdy and straightforward Croydon pattern valve is nowadays more likely to be found serving a cattle trough or a municipal allotment storage cistern, than in the home. This valve has a washered plug that moves vertically up and down within the valve body as water level in the cistern rises and falls.

The body is so shaped as to permit water to gush out, in two noisy splashing streams, through channels on each side of the plug, as water level falls. Its noisiness makes it unsuitable for domestic use.

One or other of the variations on the Portsmouth pattern ball-valve is the one most likely to be installed in the cisterns of houses built more than a decade or so ago.

The washered plug of the Portsmouth ball valve moves horizontally within the valve body. A slot in the lower part of the plug accommodates the upturned end of the float arm. As water level falls, the movement of the float arm pulls the plug away from the valve nozzle and permits water to flow into the cistern.

It used to be the practice to screw a metal or plastic 'silencer tube' into the outlet of these ball valves. It reduced the noise of refilling by introducing incoming water at a point below the level of water already in the cistern. These silencers are now forbidden by Water Authorities because of the risk of back-siphonage from the cistern into the main in the event of a failure of mains pressure.

Within the past decade a different kind of ball valve—developed at the Government's Building Research Station at Garston and known as the Garston, the BRS or the diaphragm ball-valve—has become increasingly popular.

Garston valves may be made of either brass or plastic material. Their essential feature is a large rubber diaphragm, which presses against the nylon valve nozzle to close it, as water level rises. The only moving parts are the float arm itself and the tiny plunger that it presses against the diaphragm. These moving parts are pro-

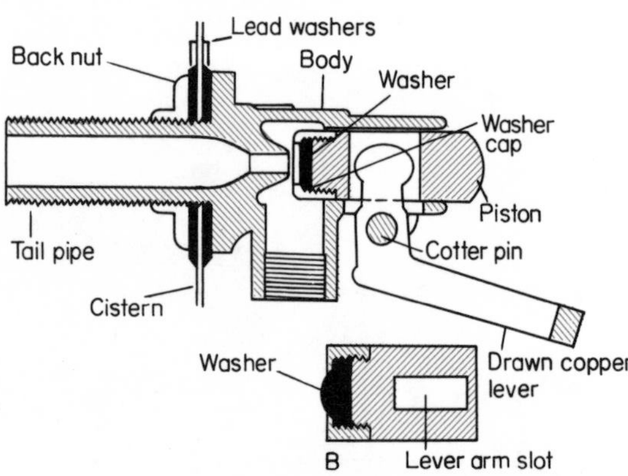

The Portsmouth pattern ball-valve is still the most common type in use. At the prompting of the float arm or lever, a brass plug into which the washer fits, moves horizontally in the valve body to open or close the valve. It used to be the practice to screw a 'silencer tube' into the outlet but this is now prohibited.

tected from water by the diaphragm and therefore cannot be affected by hard water scale or corrosion. The nylon nozzle ensures a smooth and relatively silent delivery of water.

Early models of these valves were provided with screw-in or push-on silencer tubes similar to those available for Portsmouth valves. Since these silencer tubes have been prohibited, the manufacturers have produced Garston valves with an overhead outlet incorporating a sprinkler device that delivers the water as a gentle shower, rather than as a noisy stream.

Ball-valves are described either as 'high pressure', 'low pressure' or 'full way' depending upon the size of the nozzle orifice. Valves connected directly to the rising main are normally 'high pressure'; those supplied from a storage cistern in the roof space 'low pressure'. Where, as in the case of a flushing cistern supplied from a packaged plumbing system, the main cistern is only a few feet above the level of the flushing cistern, a full-way valve may be required.

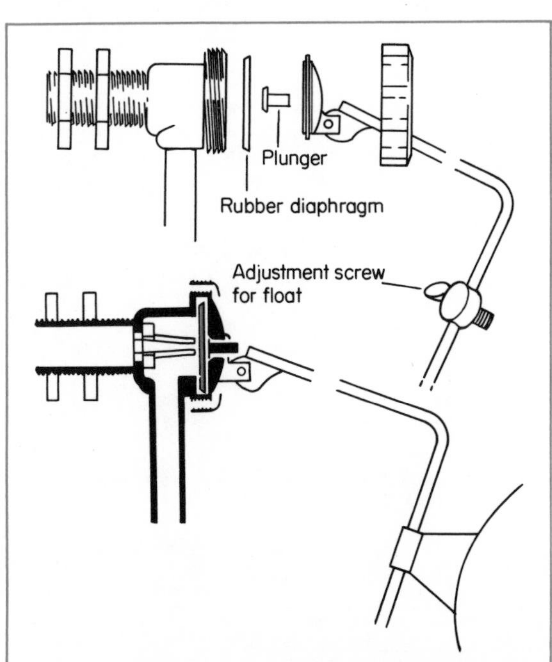

An early Garston or Diaphragm pattern ball valve. The end of the float arm pushes a small plunger against a large rubber diaphragm to close the valve. Design points are easy dismantling, moving parts protected from the water by the rubber diaphragm, detachable nylon nozzle and means of adjusting water level by moving the float up or down the vertical end section of the float arm. The silencer tube shown is now prohibited.

In some areas water pressure in the mains fluctuates considerably. A high pressure valve will deliver water unacceptably slowly during the day. On the other hand, during the night when pressure is higher, a low pressure valve will let water pass and produce a dripping overflow pipe.

The answer, under such circumstances, is to fix an equilibrium ball valve. The Portsmouth version of an equilibrium valve has a channel passing through the plug to admit water into a chamber behind it. Thus, there is equal pressure in front of and behind the valve. It is in a state of equilibrium and is activated solely by up and down movement of the float arm; not by the pressure of water trying to force it open.

These valves also eliminate the 'bounce' on the valve seating as the valve closes. This a frequent cause of water hammer and other ball-valve noise.

Until recently equilibrium valves have been available only in the Croydon and Portsmouth patterns. Now however we have the Torbeck diaphragm/equilibrium valve to which I referred earlier. This incorporates features of the Garston and the conventional equilibrium valve.

The Torbeck

The Torbeck, like a conventional equilibrium ball valve, has a water chamber—the 'servo chamber'—behind the diaphragm closing the nozzle aperture. Water flows into this chamber via the metering pin opening but is prevented from passing through into the cistern, when this is full of water, by the sealing washer on the float arm. This closes the pilot hole.

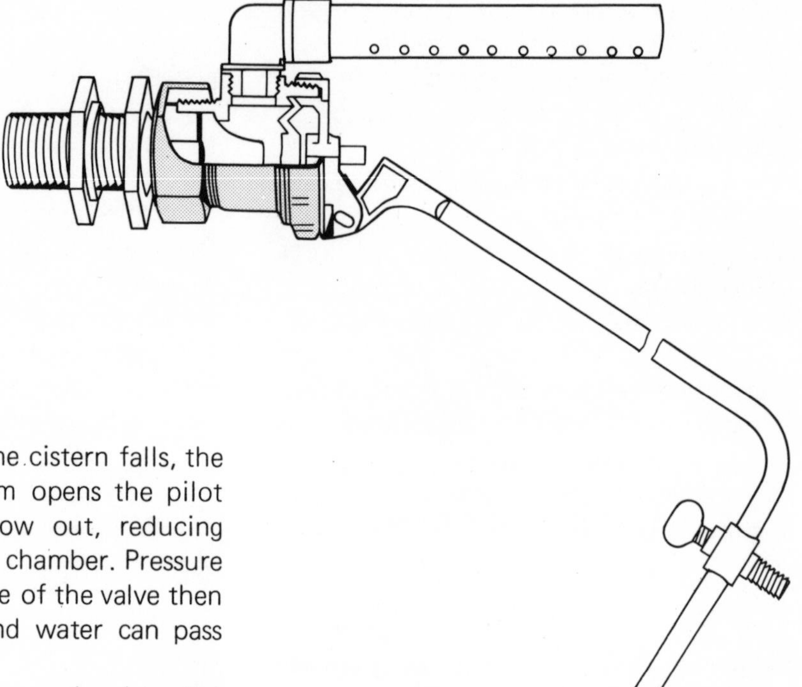

When water level in the cistern falls, the descent of the float arm opens the pilot hole and water can flow out, reducing the pressure in the servo chamber. Pressure of water on the inlet side of the valve then opens the diaphragm and water can pass through into the outlet.

The Torbeck has an overhead outlet like a modern Garston valve. This is fitted with a collapsible plastic silencer tube, which reduces the noise of water delivery but is immune to the risk of back siphonage.

The design of the valve has permitted the use of a very small float and short float arm—an important consideration when replacing the ball-valve of a lavatory flushing cistern.

A modern diaphragm ball valve by Peglers. Note the overhead outlet with spray delivery tube and the readily demountable nylon nozzle which permits rapid conversion of the ball valve from high to low pressure.

The plug of an equilibrium ball valve of this type has a channel drilled through it to permit water to pass through to a space behind the plug. Thus water pressure on either side of the plug is equal and water pressure is not continually trying to force the valve open.

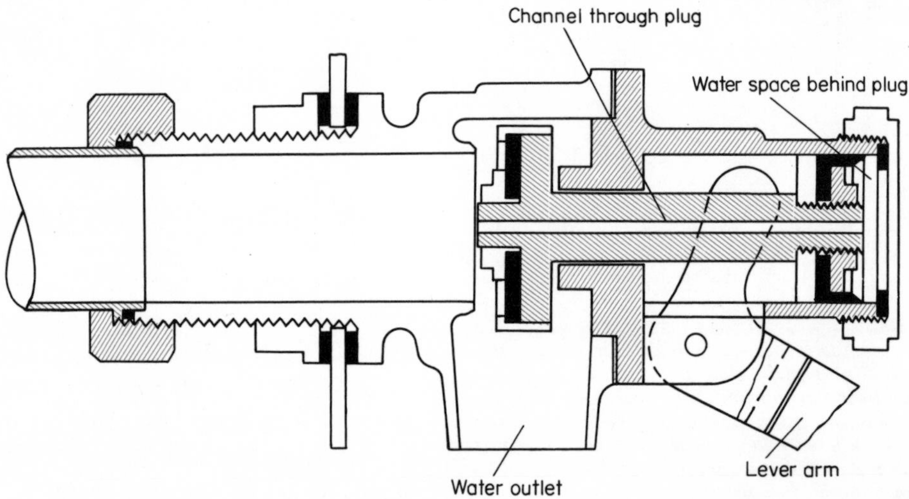

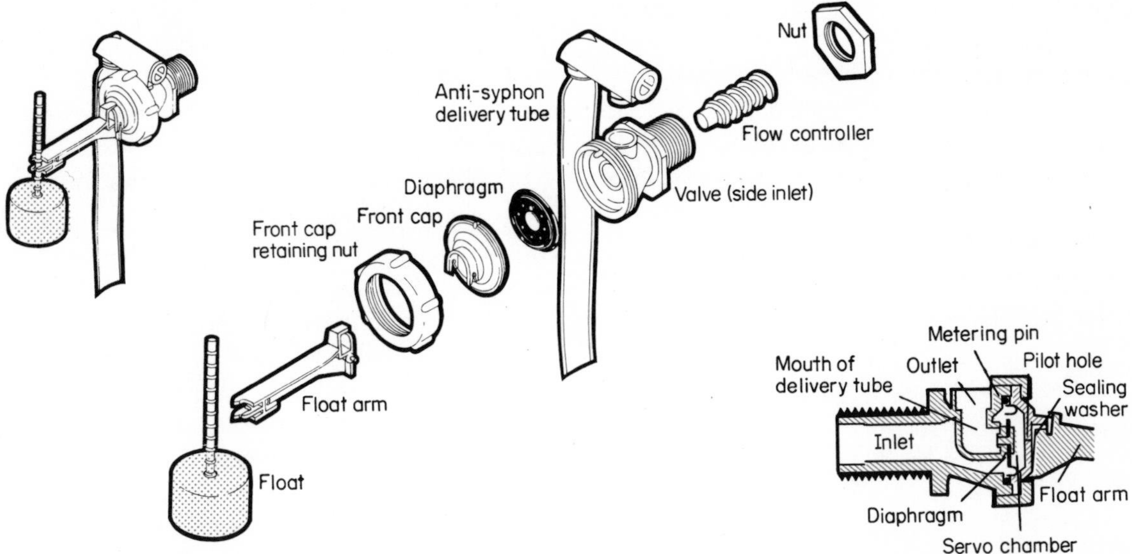

The Torbeck diaphragm/equilibrium valve embodies some of the features of the diaphragm valve and some of the conventional equilibrium valve. Water pressure is used to open and close the valve. It has been found to be efficient and silent in action.

The valve has a wide nozzle aperture, permitting rapid filling. A flow controller is provided for use where the valve is connected directly to a mains supply.

Faults in ball valves

1. Leaking valve—becomes apparent as a steady drip, or possibly a stream of water, from the cistern overflow pipe

With a Portsmouth ball valve the most probable cause is failure of the washer.

To renew the washer cut off the water supply to the ball valve. Withdraw the split pin on which the float arm pivots and remove the float arm. Insert the blade of a screw-driver into the opening from which the float arm has been withdrawn and push the plug, with its washer, out of the end of the valve body.

The plug is in two parts; though this may not be apparent on inspection. It should be possible to remove the washer retaining cap by slipping the blade of a screw-driver through the slot in the plug and then by turning the cap with a pair of pliers.

This can be extremely difficult. Don't risk damaging the plug in your efforts. If the cap cannot be unscrewed, pick out the old washer with the point of a penknife blade and force the new one under the flange of the cap. Make sure that it lies flat on its seating.

Before reassembling the valve, clean the plug with a piece of fine abrasive paper. Wrap a piece of abrasive paper round a pencil and clean the inside of the valve body in the same way. Apply a thin film of petroleum jelly to the plug to act as a lubricant and reassemble.

Other possible causes of a leaky valve are:

A leaking ball float. Renewal of the float is the answer but a temporary repair can be made by enlarging the leak and draining the water out of the ball. Replace on float arm and slip over it a small plastic bag. Secure the neck of the bag around the float arm with a piece of string.

Ball-valve wrongly adjusted for water level. Garston and Torbeck valves are provided with simple means of raising or lowering the float. To adjust the level of water in a cistern served by a Portsmouth valve, unscrew and remove the ball float. Take the float arm gently but firmly in both hands and bend the float end upwards to raise the water level; downwards to lower it.

Low pressure valve connected directly to high pressure main. Remedy is to replace with high pressure valve or, if nozzle inlet is detachable, with high pressure nozzle. Alternatively fit an equilibrium ball valve.

2. Slow refilling after water has been flushed or drawn from cistern

Possible causes of this trouble are:

High pressure valve connected to a water supply from a storage cistern. Replace with a low pressure or full-way valve or, if nozzle is detachable, with a low pressure nozzle. Alternatively fit an equilibrium ball valve.

Hard water scale impeding plug of Portsmouth valve. Dismantle valve. Clean and lubricate plug and interior of valve body as suggested above after rewashering.

Debris from main or supply pipe impeding flow through diaphragm valve. Trouble from this cause is likely to have a sudden onset. Dismantle diaphragm valve and remove debris. To dismantle a diaphragm valve turn off the water supply to the valve. It should be possible to unscrew the large knurled head that retains the valve mechanism by hand. When reassembling, screw up hand-tight only.

3. Ball valve noise

The sound of rushing water is only one, and perhaps the least disturbing, of the noises for which a ball valve may be responsible.

Other noises may include the heavy knocking of water hammer as the valve—or even the tap over the kitchen sink—closes; a roar that I have heard described as being similar to an express train entering a tunnel; or a relatively gentle humming noise that goes on and on—and on!

Water hammer results from the valve bouncing on its seating as pressure from the main attempts to force it open and the buoyancy of the float endeavours to keep it closed. Replacement with one or other of the equilibrium valves that have been described is the answer.

The other noises result from ripple formation on the surface of the water in the cistern as water flows in. These ripples shake the float up and down and to and fro. This movement is transmitted to the valve and thence to the water supply pipe which—especially if it is of copper—acts as a sounding board, to produce a noise out of all proportion to its original cause.

A stabiliser—a plastic disc or even a plastic flower pot—fixed to the float arm so as to be suspended in the water an inch or so below the float, will help. Fitting a modern Garston pattern or Torbeck valve should cure the trouble.

Another point to watch, especially when replacing a metal cold water storage cistern with a plastic one (see chapter on 'Cold Water Services and Cisterns'), is that the cold water supply pipe to the storage cistern is securely fixed to the roof timbers.

Chapter 5
The lavatory

The lavatory, by whichever name you prefer to call it—the loo, the toilet, the w.c.—may be the smallest room in the house, but it is by no means the least important. It is a room that even your least exacting guest will expect to be readily available, silent, unobtrusive and 100% efficient.

In a modern home one would expect to find two lavatories. One might be near the back door, perhaps even approached externally, readily available to the children of the household when playing in the garden. The other will be adjacent to, or perhaps within, the bathroom.

Lavatory suites are often classified loosely as 'high level' or 'low level' but there are a great many more variations.

Let us take the flushing cistern first. British cisterns are constructed to give a two gallon flush. In the building trade they are often called 'water waste preventers' or WWPs for short. There are, at the present time, two different kinds of flushing cistern in common use.

Cast iron cisterns

The older pattern is the traditional high level cistern, sometimes known as the 'bell' or 'pull and let go' type. Rapidly becoming obsolete, it is still to be found, particularly in outside lavatories, in older houses.

This kind of cistern is usually made of cast iron. It has a well in its base in which stands a heavy iron 'bell'. A stand-pipe outlet to the flush pipe rises from the base of the cistern, inside the bell, to terminate open-ended an inch or so above normal water level.

The cistern is flushed by raising the bell, usually by pulling a chain, and then releasing it. The bell's conical shape forces the water within it upwards and over the rim of the stand-pipe as it descends. The falling water mixes with air in the flush-pipe, creating a partial vacuum and thereby starting the siphonic action that empties the cistern.

The bell has metal lugs on its base that permit water to pass under the rim and up to the stand-pipe outlet, once the siphonic action has started.

Strong, hard wearing and reliable as these cisterns usually are, they have a number of disadvantages. They are noisy in action. Noise results from the clank of the bell as it descends, the rush of water from high level and, since they are usually found connected direct to the main, the incoming rush of water as the cistern refills. At their best they are hardly objects of beauty. All too often they are to be found dripping with condensation, with flaking paint and large patches of rust.

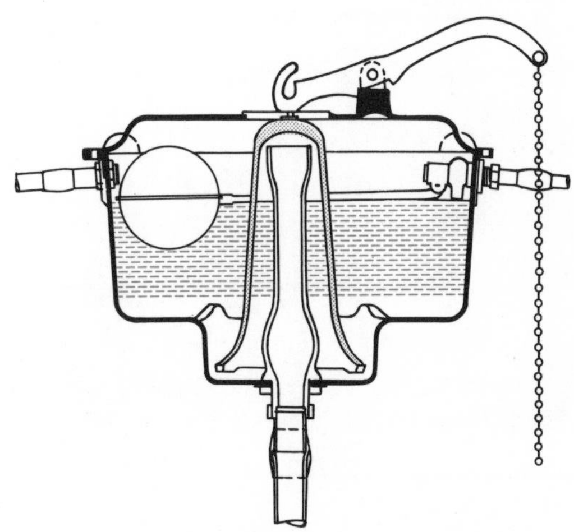

A bell or Burlington pattern flushing cistern operates as a result of the bell, when descending, forcing water over the lip of the standpipe into the flush pipe. Descending water mixes with air in the flush pipe to create the partial vacuum on which siphonic action depends.

Plastic cisterns

These objections are, to a greater or less extent—depending upon the particular appliance—overcome by the more modern 'direct action' cistern. These are sometimes described as low level cisterns though they can be, and frequently are, installed at high level.

This type of cistern is usually made either of ceramic or plastic material. Most have a flat base, though some well-bottomed models are manufactured to facilitate the replacement of old bell cisterns. The flushing mechanism consists of a stand-pipe rising to above water level and then bending over and opening out to form an open-based dome. When the flushing lever is operated a disc is raised within this dome to throw water over the inverted U at the top of the siphon into the flush pipe, thus starting the siphonic action.

The disc has a hole, or holes in it to permit water to pass through freely once the siphonic action has begun. As the disc is raised these holes are closed by a valve--usually nowadays a simple plastic flap.

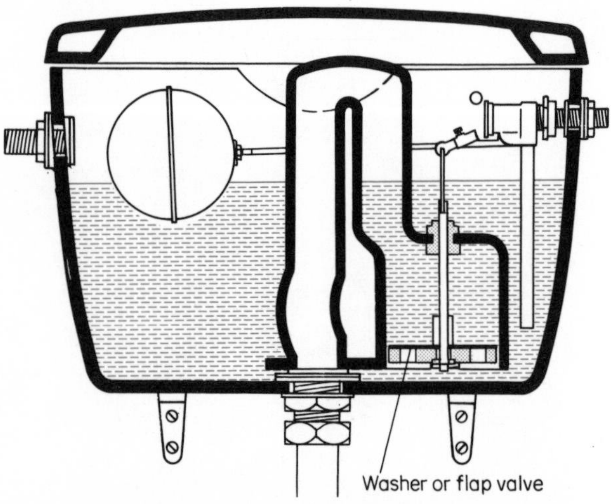

Operating the lever of a direct action flushing cistern raises a plate within the dome of the siphon. This throws water over the inverted U of the siphoning mechanism to start the flushing action. The plate has holes in it to permit water to pass through freely during the flush. These holes are closed when the plate is raised by means of a flap valve or diaphragm.

The slim-line cistern

A recent development, of considerable interest to the d.i.y enthusiast, has been the advent of the slim-line flushing cistern or 'flush panel'. Space saving in a new bathroom or lavatory, it also simplifies the conversion of a high level to a low level suite.

The lavatory pan of a conventional low level suite is situated a couple of inches or so further from the wall than that of a high level suite. This is to accommodate the flushing cistern and to enable the seat to be raised when required.

In the past, the conversion of a high level to a low level suite has meant moving the lavatory pan forward and extending the branch drain or branch soil-pipe to which it is connected—a somewhat daunting task! The slim line cistern or flush-panel has changed all that. It is usually possible, using one of these cisterns, to convert from high level to low level without the need to change the position of the pan.

Another new development, likely to become more widely known as the need for water conservation increases, is the dual-flush cistern. On a great many occasions a full two gallon flush is not needed after the use of the lavatory. If the operating handle of a dual-flush cistern is depressed and immediately released, the cistern gives a one gallon flush only. For a full flush the lever must be held down for a few moments.

Types of pan

There are three basic designs of modern lavatory pan: the wash-down, the single-

A wash-down low level lavatory suite depends upon the weight and momentum of the 2 gal flush to cleanse the pan. A conventional suite of this kind has the pan several inches further from the wall than does the equivalent high level suite. This is to permit the flushing cistern to be accommodated. (Courtesy Fordham Pressings Ltd)

The use of a slim-line flushing cistern such as the Fordham flush panel usually makes it possible to convert a high level lavatory suite to low level operation without moving the position of the pan. (Courtesy Fordham Pressings Ltd)

trap siphonic and the double-trap siphonic pan.

The *wash-down pan* is the basic lavatory pan with which everyone is familiar. Cleaning of the pan and its recharging with water depend upon the weight and momentum of the two-gallon flush entering at the back of the pan and via the flushing rim.

The other two types of pan depend, at least partially, on the weight of the atmosphere, upon siphonic action, for their effectiveness. They permit the use of 'close coupled' lavatory suites in which pan and cistern comprise one unit, without even the short flush pipe of a low level suite.

Single-trap siphonic pans are so designed, either by means of a constriction or a bend in the outlet, to ensure that the outlet pipe fills with water as the pan is flushed. This results in the escaping water pushing the air in the pipe in front of it and producing a partial vacuum—and siphonic action. With these appliances, water level will rise slightly in the pan as the flush first operates and will then empty rapidly, perhaps with a gurgle as the siphon is broken.

Double-trap siphonic suites operate on an ingenious principle that ensures exceptionally silent and effective action. As the first flow of water passes from the cistern to the pan, air is aspirated—by means of a pressure reducing device—from the space between the two traps. The partial vacuum produced ensures that atmospheric pressure pushes out the contents of the pan. The flushing water is required only to set this process in motion and to recharge the pan.

With a properly designed and installed

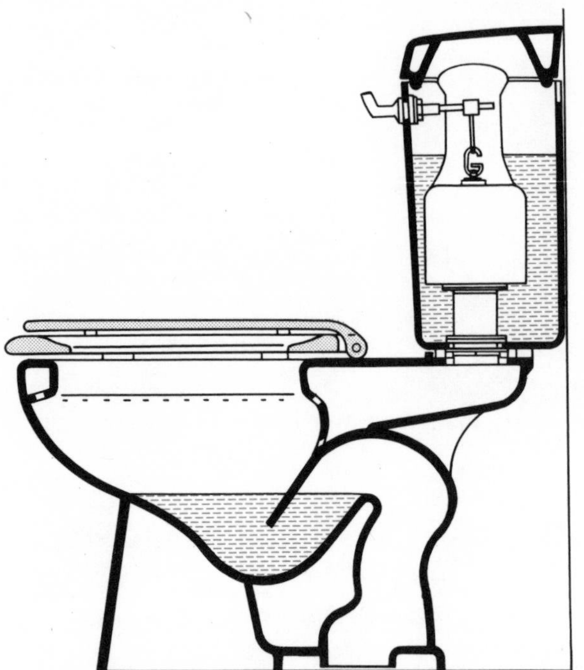

Siphonic lavatory suites can be 'close coupled' — no visual flush pipe is needed. At the start of the flush of a single trap siphonic suite, such as the Twyfords suite illustrated, water rises in the pan. It then rapidly discharges over the outlet of the trap, completely filling the constricted part of the outlet. This creates a strong siphonic action.

double trap siphonic suite the water level in the pan should fall visibly ***before*** water descends from the flushing rim.

Siphonic suites, particularly double-trap siphonic suites, are silent in operation, permit neat close-coupled construction and a large water area. They are especially to be recommended where, as in a lavatory opening from an entrance lobby, silent unobtrusive operation is of first importance.

The only disadvantage of this kind of suite arises from possible misuse. A foreign body, such as a plastic toy or a small cleaning brush, flushed from a siphonic suite can create a blockage that is particularly difficult to clear.

When contemplating the replacement of a washdown suite with a double trap siphonic type, remember that the latter will project further from the wall.

Replacing a cracked pan

Renewing a cracked or leaking wash-down lavatory pan is a job that has been successfully undertaken by many d.i.y enthusiasts. Removing the old pan is likely to present the greatest difficulty.

This is easy enough where the lavatory is an upstairs one and has a wooden boarded floor. The pan, in this case, will be screwed to the wooden floor and connected to the branch soil-pipe by means of a mastic or a patent push-on plastic joint. The flush-pipe connection must be disconnected and the floor screws removed. The pan can then be pulled forward and disposed of.

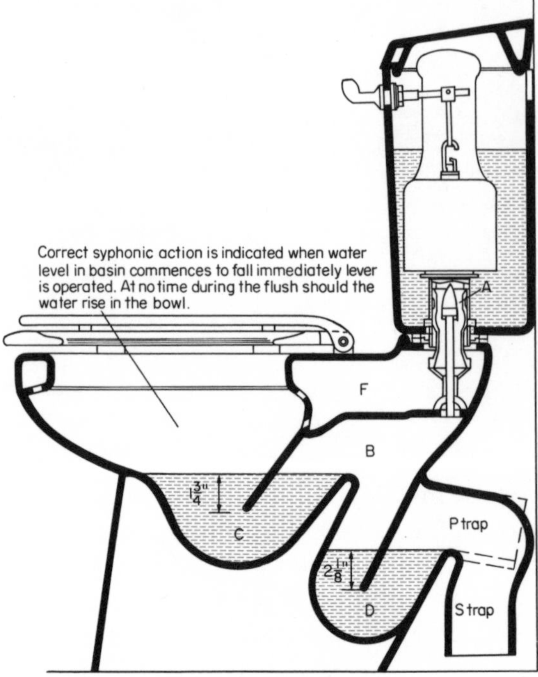

Double-trap siphonic lavatory suites are particularly silent and effective in action. In the Twyfords model illustrated, water flowing from the flushing cistern passes over the pressure reducing fitment A. This aspirates air from the chamber B and siphonic action draws the contents of the pan through the sealed traps C and D. The sides of the pan are thoroughly washed and cleansed by streams of water from the perforated rim E. After flushing, the pan is filled and the traps resealed by the emptying of afterflush chamber F.

Correct siphonic action is indicated by water level in the pan falling immediately the flushing lever is operated — before water appears from the flushing rim.

Greater difficulty arises where the pan is on a solid ground floor and is connected to the branch underground drain by means of a cement joint.

Tackle its removal this way:

Disconnect the flush pipe. Deliberately break the outlet from the old pan with a hammer, just behind the trap. Remove the retaining screws, if any, and pull the front part of the pan forward. If, as is likely, it is cemented to the floor, you will need a cold chisel to prise it away from its base.

You will now be left with the broken outlet of the lavatory pan protruding from the drain socket. Stuff a wad of newspaper into the drain socket to prevent pieces of broken pipe and cement falling into the drain. Then tackle the outlet with a cold chisel and hammer.

Work carefully and systematically, keeping the blade of the chisel pointing towards the centre of the pipe. Try to break down the lavatory outlet to the base of the socket at one point. You will then find that the remainder comes out fairly easily. Clear away the jointing material in the same way.

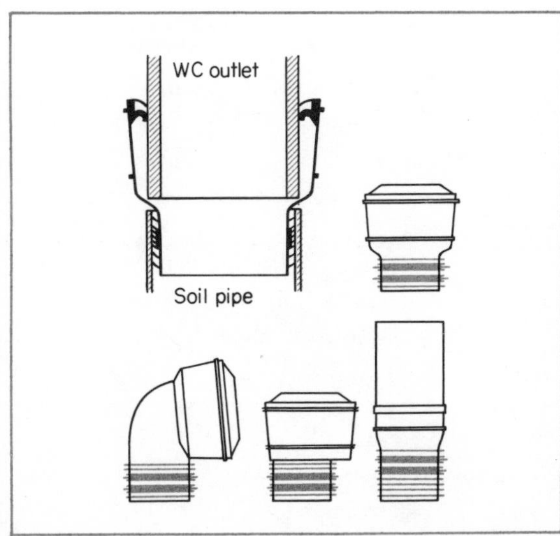

Push-on plastic w.c. connectors make the connection of the lavatory pan to the branch drain or soil pipe a simple task.

Try not to break the drain socket but, if you accidentally do so, don't despair. Modern plastic push-on drain connectors can be used, without a socket, directly into the drain pipe.

Do ***not*** set the new lavatory pan on a base of wet cement. It has been established that setting cement can set up tensions that can lead to early pan damage.

Place the pan in position and mark through the screw-holes with a ball point pen refill on to the floor. Remove the pan. Drill and plug the points that you have marked. Replace the pan and screw down - not too hard - using lead washers to protect the pan from the screw heads.

I have suggested using a plastic push-on connector for the drain connection. An alternative is to bind a couple of turns of waterproof building tape round the pan outlet and caulk down hard into the drain socket. Fill in the space between outlet and socket with a non-setting mastic filler and complete the joint with another couple of turns of waterproof tape.

Don't forget to remove the wad of paper from the drain pipe before placing the pan in position!

Faults in lavatory installations

1. Failure to flush when lever is operated

This is a common and embarrassing fault. Check that the water level in the cistern is correct: about ½ in below the overflow pipe.

If the water level is correct then the fault is almost certainly due to the failure of the flap valve that closes the holes in the disc within the siphon when this disc is raised.

You must remove the siphon to renew this valve. Tie up the ball-valve to prevent water flowing in and flush cistern to empty. Disconnect the flush pipe. The siphon can then usually be withdrawn after unscrewing the large nut immediately beneath the cistern.

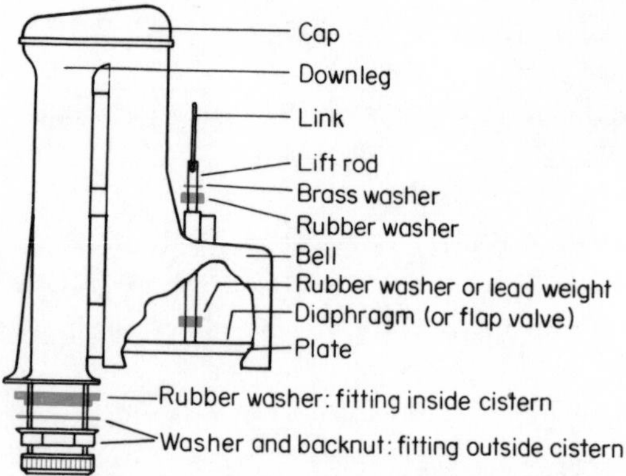

A view of the siphon removed from the flushing cistern.

Have a new valve ready for replacement. These plastic 'flap-valves' or 'siphon washers' can be obtained from any builders' merchant. You should get a valve large enough to cover the disc and touch the sides of the siphon dome.

2. Cistern fills too slowly

After flushing, the cistern should be ready for use again within two minutes. If it is not, it is probable that a high pressure ball valve has been fitted to a low pressure water supply. Alternatively the valve may be clogged with scale or, with a Garston pattern valve, debris from the main or main storage tank may be clogging the nozzle outlet (see 'Faults in Ball Valves in Chapter 4).

3. Cistern flushes but fails to cleanse pan

Check that the flush pipe connects to the pan inlet squarely and that the inlet is unobstructed. Check, with the fingers or a mirror, that the flushing rim is clear. Check, with a spirit level, that the pan is set dead level.

When flushed, water should run equally round each side of the flushing rim to meet at the centre. There should be no whirl-pool effect.

4. Double-trap siphonic suite fails to siphon out when flush is operated

This is usually due to obstruction of the pressure reducing device with jointing material or debris from the cistern.

5. Leakage from joint between lavatory pan outlet and soil-pipe or drain socket

Renew this joint as suggested in instructions for renewing a defective lavatory pan.

6. Condensation on flushing cistern—perhaps giving the impression that the cistern has become porous

The real answer to this problem lies in better ventilation and the provision of a radiant heat source. Where the lavatory is in the bathroom avoid drip-drying clothes over the bath. Always run an inch or so of cold water into the bath before turning on the hot tap.

Intractable cases may be improved by lining the inside of the cistern with strips of expanded polystyrene—as used for insulation under wall paper. Dry the cistern thoroughly and use an epoxy resin adhesive. Do not refill the cistern until the adhesive is thoroughly set.

7. Noise from the lavatory suite

For noise in filling see 'Faults in Ball Valves', Chapter 4. The most probable cause is failure of the ball valve washer. For noise when in use, bear in mind that a low level suite is quieter than a high level one and that the quietest of all is the close coupled double trap siphonic suite. Make sure that there is a mastic, not a cement, joint between the pan outlet and the branch drain or soil pipe.

Noise from an upstairs lavatory may be reduced by raising a floor board and running fine sand or vermiculate chips on to the ceiling of the room below.

Chapter 6
Baths and showers

Baths come in all shapes and sizes these days. Some appear to have been specially designed to meet the suggestion made, in the interests of economy, by a Government Minister a few years ago that married couples should, 'share a bath'. Others seem to be much too splendid examples of domestic architecture to be relegated to the bathroom!

You can be quite confident that, whether you live in a cottage, a suburban semi or a mansion, you will find in the showrooms and the manufacturers' glossy catalogues, the size and shape of bath that you need. The price, which will depend to a large extent upon the material of which the bath is made, may be another matter.

The traditional material is enamelled cast iron. Enamelled cast iron baths are strong, tough, hard wearing and extremely heavy and expensive! As cast iron is a good conductor of heat they have the added disadvantage of robbing the bath water of its warmth. This is of greater importance nowadays than it was in the past when unlimited supplies of hot water could be cheaply obtained.

Enamelled pressed steel baths are made of material similar to that used in modern slim-line pressed steel hot water radiators. They are considerably cheaper than cast iron baths and, once installed, can be relied upon to give years of trouble-free service. They are, however, much more liable to accidental damage in storage, delivery and installation.

My choice nowadays, whether for d.i.y or professional installation, would be one of the modern acrylic plastic baths. These are tough, but extremely light and easily handled. They can therefore be installed by one man working alone. They are available in a variety of colours. Slight surface scratches can be polished out and, of course, surface damage will not lead to the corrosion to which the other kinds of bath are prone.

Acrylic plastic material is a poor conductor of heat. Consequently baths of this material tend to retain the heat of the water and to be comfortable in use.

They can be damaged by excessive heat however. A burning cigarette rested, even briefly, on the edge of the bath can do irreversible damage. For the same reason,

(a) A metal cradle, which speeds and simplifies installation whilst affording a base for fixing side and end panels is a standard extra of all Cleopatra range acrylic baths.

1. *The universal wallfixing bracket can be used under, at the side of, or over the lip of the bath.*
2. *Simple screw adjustments at five points compensate for irregular floor levels.*

(b) Details of the cradle for Armitage acrylic baths.

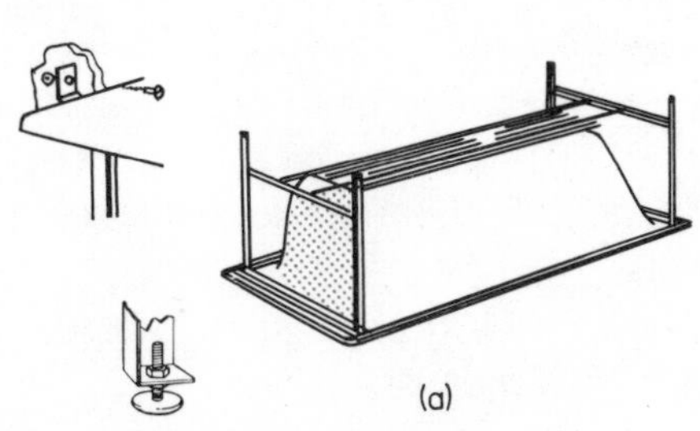

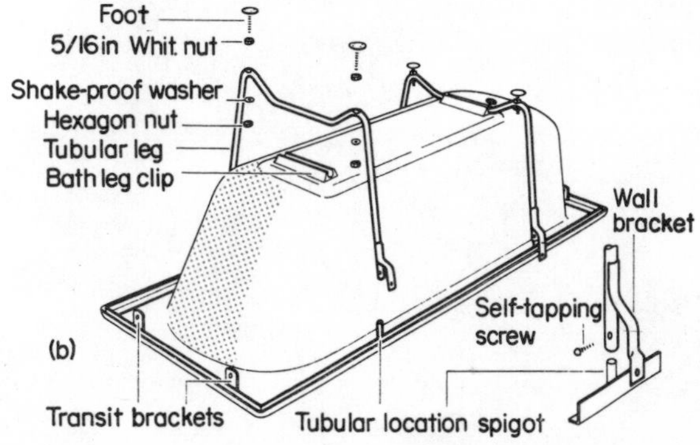

keep the flame of any blow torch that you may be using for pipe jointing, well away from the plastic material.

There was another snag about some of the earlier pioneer models. Being less rigid than metal baths, they tended to sag and creak as they were filled with water and, even more disconcertingly, as the bather stepped in.

This drawback has been overome by the manufacturers providing substantial metal or wood cradles and means of securing them to the bathroom wall. Since these baths are as new to many professional plumbers as they are to the d.i.y man, full and detailed instructions are provided with each bath sold. These instructions should be followed exactly.

The space behind the bath will be very limited and difficult to work in. As much as possible should be done before the bath is moved into position. Fix the taps or mixer as described in Chapter 4. Have the hot and cold water pipe lines in position ready to be connected to the tails of the taps by means of a compression joint incorporating a tap connector (see Chapter 11).

The trap should also be connected to the waste pipe in advance and the waste outlet of the bath bedded into the hole provided for it in a non-setting mastic such as 'Plumbers Mait'. All that will then need to be done when the bath is moved into position is to tighten up the nut connecting the waste outlet to the trap and to connect the taps and overflow pipe. Be sure to do the final operations in logical order. Connect up the further tap first, then the overflow and finally the nearer tap.

With plastic baths it is wise to use a plastic trap and waste pipe. The plastic material of the bath will move slightly as a result of expansion when filled with hot water. A rigidly fixed metal trap and waste could damage the bath plastic.

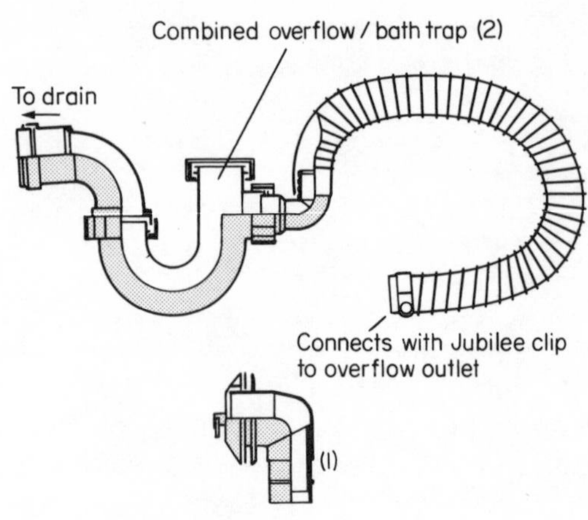

The bent bath overflow outlet 1 connects by means of a jubilee clip to the flexible overflow pipe of the combined overflow/bath trap 2

Finally, of course, you must seal off the gap between the edge of the bath and the wall. Again, a non-setting mastic is best for this purpose. There are a number of these on the market.

Faults in baths

1. Blocked waste outlet—water fails to run away when waste plug is pulled out.

Use a 'force cup' or 'sink waste plunger' to clear blockage in the trap or waste pipe. This consists of a hemisphere of rubber or plastic mounted on a wooden handle. It is a basic plumbing tool that every householder should possess.

Place the rubber hemisphere over the waste outlet. With the other hand hold a damp cloth firmly against the overflow outlet. Plunge down sharply several times with the wooden handle.

Since water cannot be compressed, the force of the plunger action will be transmitted to the obstruction to dislodge it. The overflow outlet must be blocked to prevent this force from being dissipated up the overflow pipe.

A very slow flow of water from the waste outlet indicates a partial blockage.

Check that the waste outlet grid is not obstructed with hair or other debris. If the trouble persists after this has been cleared use one of the proprietary chemical drain cleaners that you can buy at any household store.

These drain cleaners have a caustic soda base and are potentially very dangerous. Read the makers' instructions and follow them carefully. Keep the tin well away from children.

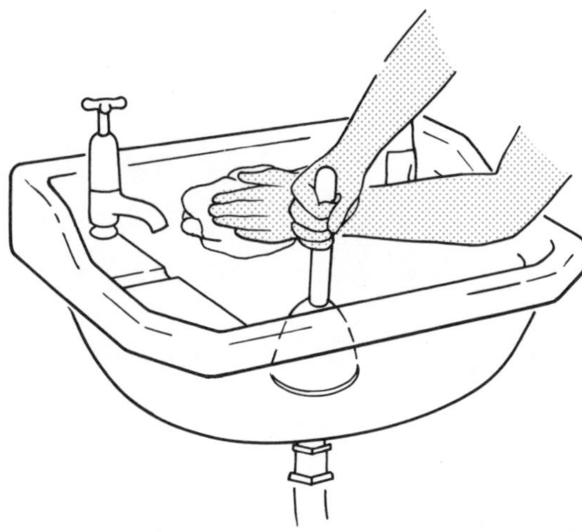

To clear a blockage in a sink, basin or bath waste hold a dampened cloth firmly over the overflow outlet while plunging the waste outlet with a force cup or plunger.

2. Stains in bath below bath taps.

Proprietary bath cleaners such as Jenolite are usually very effective in removing these. An old fashioned remedy, that I have known to succeed where modern ones have failed, is to mix up a paste with hydrogen peroxide and cream of tartar. Apply to the stains. Leave overnight and wipe off in the morning.

Stains of this kind indicate a dripping tap. Change washer as suggested in a previous chapter.

3. Enamel of cast iron or pressed steel bath worn and in poor condition

Bath renovating preparations of various kinds are advertised in all do-it-yourself publications.

An essential prerequisite of success with these preparations is thorough cleansing of the bath surfaces. Wash with a very dilute solution of hydrochloric acid and rinse. Apply a fine abrasive to the whole

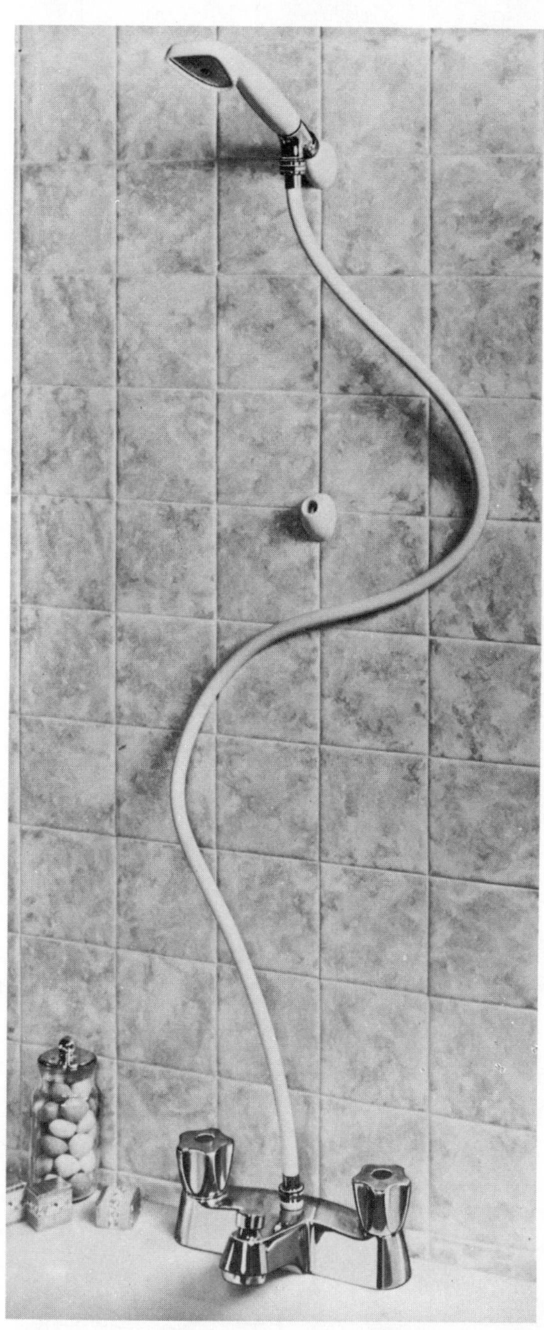

Bath/shower mixer made by Barking Brassware Company. This permits the mixed hot and cold water to be diverted upwards to the shower sprinkler at the flick of a switch.

bath surface. Wash again in very dilute hydrochloric acid and rinse with a solution of washing soda. Rinse again with plain water and dry thoroughly.

A shower cabinet may be installed on a bathroom, in a bedroom, on a landing or even in the cupboard under the stairs. (Courtesy Twyfords Ltd).

Showers

There's a great deal to be said for a shower. A shower uses less water and takes less time than a sit-down bath. It has been estimated that five or six satisfying showers can be obtained for the same amount of hot water that would be needed for one bath. Even with quite a small family a morning sit-down bath for each member would be a physical, as well as an economic, impossibility in most homes. It is by no means impossible where a shower is installed.

Showers are more hygienic—the bather is not sitting in his own dirty water—and mean less work for the housewife cleaning up afterwards. Elderly people, who may be incapable or afraid of getting into a sit-down bath, can step into a shower cubicle without trouble or anxiety.

Showers can be installed as an adjunct to a conventional bath or as a separate amenity in their own shower cabinet. They are particularly useful where older properties are being brought up to date and there is not space for a bathroom. A shower cubicle can be accommodated on a landing, in a bedroom or even in the cupboard under the stairs.

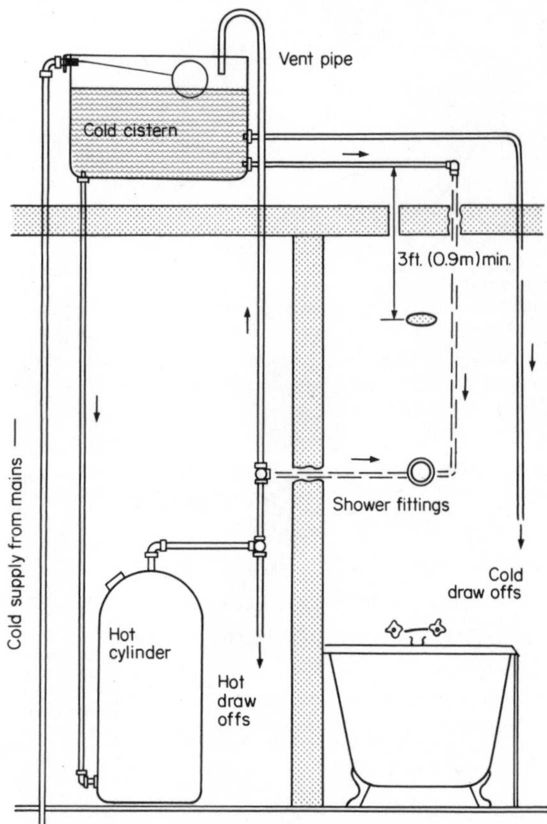

The essential design requirements for a conventional shower are:

1. *Hot and cold water supplies under equal pressure from a storage cistern.*
2. *A minimum head of 3ft but 5ft is to be preferred.*
3. *A cold water supply taken direct from the cold water storage cistern.*

There are certain minimum plumbing design requirements that must be met if a shower is to be both effective and safe.

The hot and cold water supplies to the shower must be under equal pressure. Where hot water supply is from a cylinder storage system, water pressure on the hot side of the shower will derive from the cold water storage cistern. The cold supply must therefore also be taken from this cistern. It is both illegal and impracticable to take the cold supply to a shower installation of this kind direct from the main.

Then again, pressure must be adequate. Pressure at the shower sprinkler depends upon the vertical distance between the sprinkler and the base of the cold water storage cistern. The absolute minimum vertical distance or 'head' is 3 ft. This will give an effective shower only if the pipe runs are very short and free of bends. Best results will be obtained if the 'head' between shower sprinkler and cistern base is 5 ft or more.

It should be noted that the level of the hot water cylinder relative to the shower is

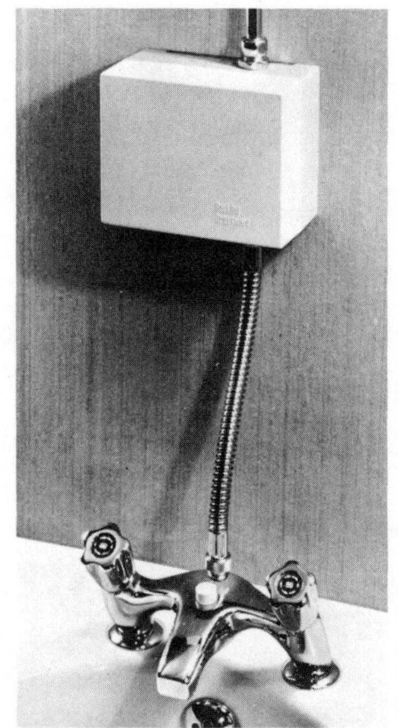

The Flomatic shower booster pump operates on a flow switch and permits shower installation even where the vertical distance between the shower sprinkler and the surface of the water in the storage cistern is no more than 200 mm (8 in) (Barking Brassware Co.)

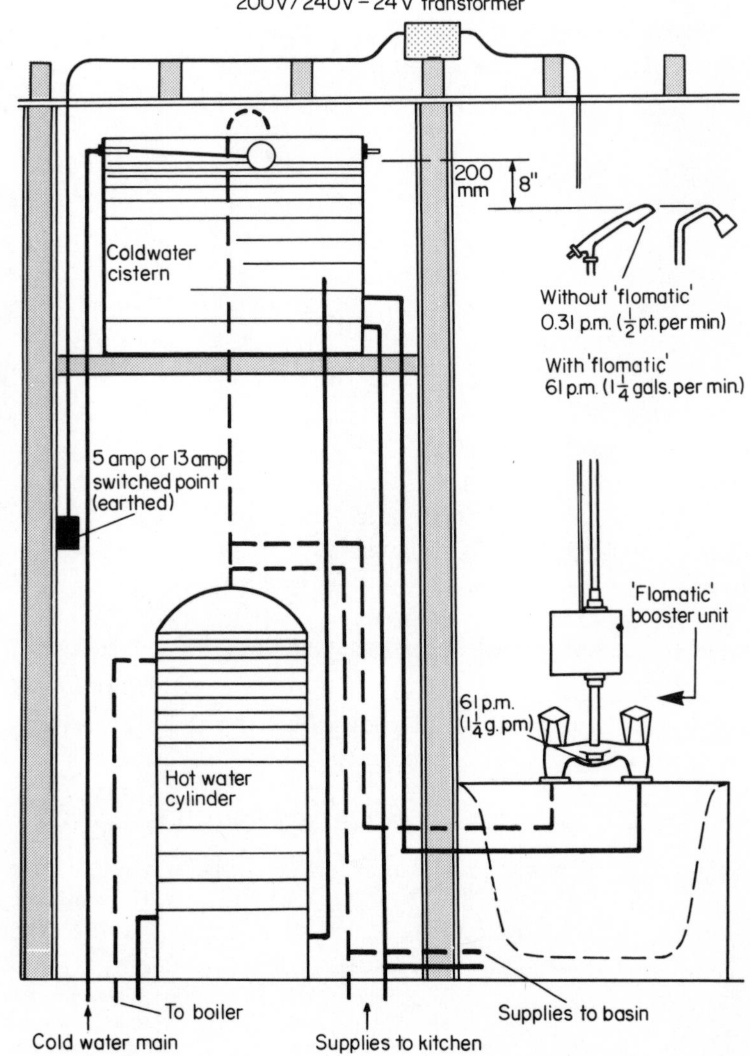

immaterial. It can be above, below or on the same level as the shower. It is the level of the ***cold water storage cistern*** that matters.

Another important design point is that the cold water supply to the shower should be taken direct from the storage cistern. It should not be a branch taken from a pipeline supplying other draw-off points.

This is a safety measure. If the cold supply were taken from such a branch pipe-line then the flushing of a lavatory or the turning on of a cold basin tap would reduce cold water flow to the shower. Dangerous scalding could result from the sudden rise in water temperature.

For a similar reason it is best to take the hot water supply in a separate pipe-line from the hot water storage cylinder. This is however a little less important A sudden drop in temperature may cause discomfort, but is hardly likely to be dangerous, to the bather.

If your plumbing system is so designed that it just is not possible for you to meet the minimum requirements that I have set out, it may still be possible for you to install a shower.

There are, nowadays, flow-operated electric shower pumps on the market that can be used to boost pressure where the minimum 3 ft head is not available. These add to the cost of installation but they can be very useful in a flat or maisonette where the level of the storage cistern cannot be raised.

Some manufacturers of multipoint gas instantaneous heaters provide the means of supplying a shower from their appliances. As these water heaters are usually fed direct from the main, cold water from the main may be permitted to mix with hot water from the appliance. A patent anti-scald valve is an essential feature of installations of this kind.

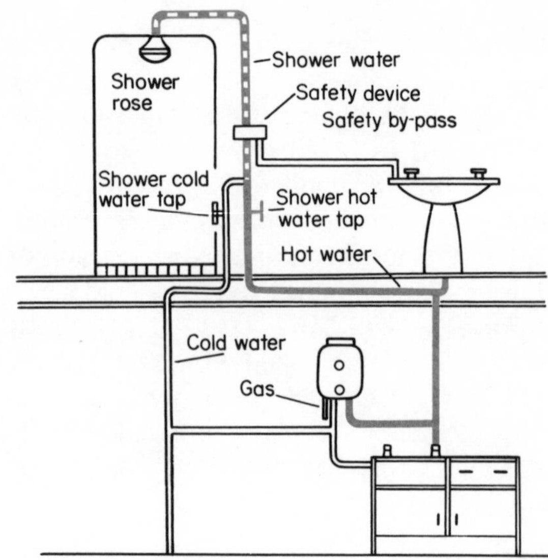

A shower can sometimes be connected directly to an instantaneous gas hot water system without the need for a cold water storage cistern.

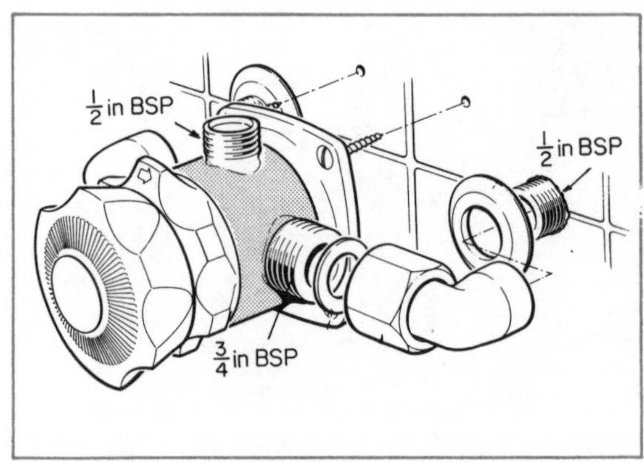

A Bourner's manual mixing valve can be fixed as shown

There are also a number of electrically heated instantaneous showers on the market. These need only connecting to the rising main and to a suitable electricity supply. The simplicity with which these appliances can be installed in practically any situation has made them increasingly popular in recent years. It should be said however that their rate of delivery of warm water is considerably less than that from a conventional shower.

Mixing valves

All conventional showers are provided with some kind of mixing valve to enable the user to vary the temperature of the water at will. The simplest kind of mixing valve consists of the two bath taps. Rubber push-on connectors can be used to join these taps to a shower attachment above the bath. Water temperature and flow are adjusted by opening the taps until the shower water is at the required temperature.

An improvement on this arrangement is to be found in the combined bath mixer/ shower from which water at the required temperature can flow into the bath from the spout of the mixer or, at the flick of a control knob, be diverted up to the shower fitting.

Most independent showers have a single manual mixing valve which can be turned to vary the temperature and, in some instances, the flow of the water.

Yet another refinement is the thermostatic mixing valve. A thermostatic valve will maintain water temperature at a constant level despite fluctuating pressures in either the hot or the cold supply. It can therefore cancel the otherwise essential requirement that every shower must have its own independent cold water supply.

It is important that the limitations of thermostatic valves should be appreciated. They cannot ***increase*** pressure in either the hot or the cold water supply. They will simply reduce the pressure on one side of the valve to match that on the other. If, for instance, cold water pressure should drop in the thermostatic mixing valve of a shower already operating on minimum head, then the shower would simply dry up until pressure was restored.

Both bath and independent showers must be provided with plastic curtains or a glass or plastic screen to prevent spillage. Independent showers have a foot tray which may be made of ceramic, enamelled steel or plastic material. The waste outlet from a shower tray is connected to the waste pipe in the same way as the outlet from a bath.

Faults in shower installations

1. Impossibility of obtaining a shower at a comfortable temperature. Shower runs cold and then, after adjustment of the taps or valve, suddenly hot

This is usually the result of having the hot supply under pressure from a storage cistern and the cold supply direct from the main. The remedy is to take the cold supply from the cistern supplying the hot water cylinder.

2. Water descends from sprinkler in a feeble stream instead of a shower

This results from insufficient water pressure. The cheapest remedy is to raise the level of the cold water storage cistern, if necessary building a platform for it within the roof space. If this cannot be done consider the possibility of providing an electric shower pump.

3. Distributing holes in shower sprinkler blocked, giving reduced flow

This usually results from scale formation within the shower head. Dismantle and clean out. Consider the possibility of installing a mains water softener (see chapter 9).

Chapter 7 Sinks, basins and bidets

Sinks, basins and bidets, despite their very different uses, present similar plumbing problems and can conveniently be considered together.

Sinks

In the '30s and '40s, the glazed ceramic Belfast sink with its wooden draining board, was regarded as the epitome of surburban luxury. It was certainly a tremendous improvement on the shallow 'London pattern' sink that had preceded it.

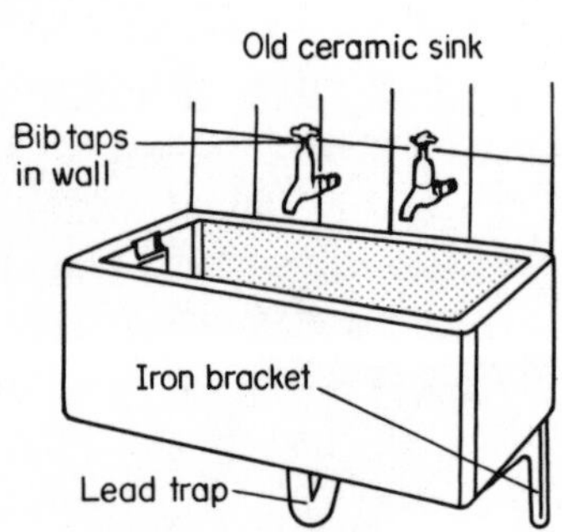

(a)

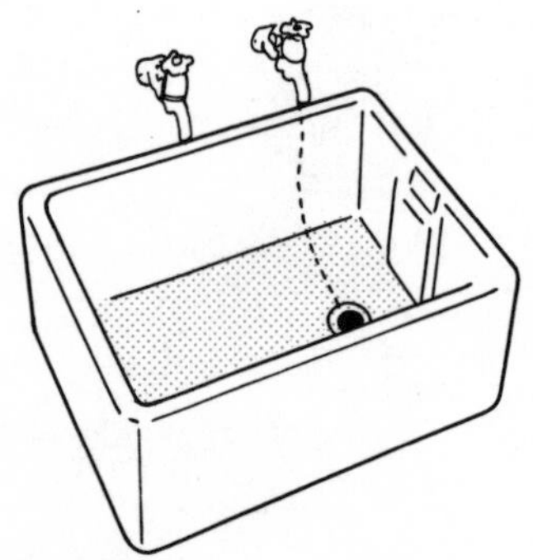

There are still plenty of Belfast sinks about. They are usually supported by strong brackets built into the wall. They have a weir overflow and are usually supplied with water from bib-taps projecting from the glazed tiles of the wall behind them. Their hard, unyielding base spells certain destruction to any item of crockery accidentally dropped into them.

Nowadays they are being replaced by sink units incorporating a pressed enamelled steel or—more probably—a stainless steel sink and drainer. Units of this kind are the standard fitting in new homes.

Enamelled steel sinks are obtainable in a number of attractive colours to match the kitchen decor but they have the disadvantage that the enamel can be chipped, and the sink permanently ruined, by accidental damage. Unless a plastic can be produced capable of standing up to the very heavy use—and misuse—to which sinks are subjected, it seems likely that stainless steel will remain the most popular kitchen sink material for years to come.

(a) Ceramic 'Belfast' pattern sinks are often fixed to the kitchen wall with strong cantilever brackets which must be removed or cut back flush with the wall when replacing with a modern sink unit. Water supply is usually from bib taps protruding from the tiled wall above the sink.
(b) A modern sink unit with 'pillar type' sink mixer.

A waste trap
B waste pipe passing through wall
C elbow connector joints two straight lengths of waste pipe
D pipe clip supporting waste pipe
E gulley

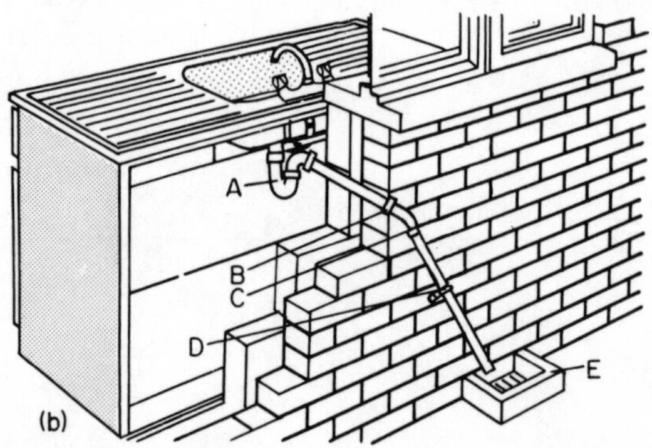

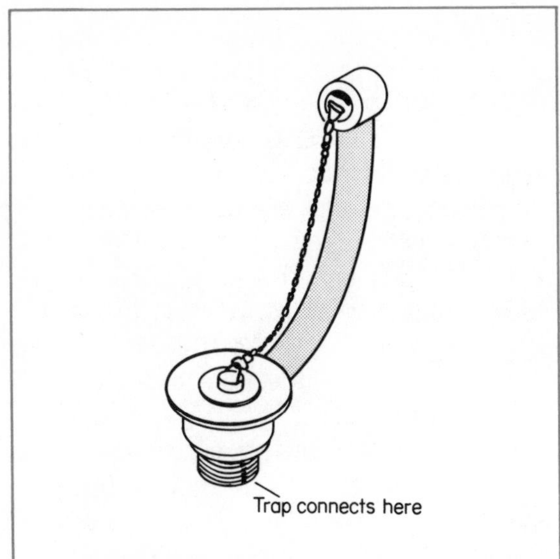

The combined waste and overflow of a modern stainless steel sink resembles that of a bath.

Stainless steel sinks may be provided with double or single drainers. Some are made with double sinks to facilitate washing up and hot rinsing or, alternatively, to permit washing up and food preparation to proceed at the same time. Most are of the traditional rectangular shape but recently smaller round sinks have become popular. These make it possible to dispense with the washing up bowl.

Water supply is provided by pillar taps, or by a sink mixer, fitted into holes provided at the rear of the sink (see Chapter 4). Some stainless steel sinks are provided with a built-in overflow but the tendency nowadays is to manufacture them with an overflow outlet only. This is connected to the waste outlet by means of a flexible pipe similar to that used for bath overflows.

All sinks must have a trapped outlet; either the traditional U bend or the more attractive looking bottle trap. The trap seal may be 50 mm (2 in) if discharging over a drain gully or connected to a two-pipe drainage system. If the waste outlet is connected to a single stack drainage system (see Chapter 8) it must have a seal of 75 mm (3 in).

The waste disposal unit

Another labour saving device (or status symbol) that can be attached to the modern sink is the sink waste disposal unit; known as a garbage grinder in the USA. Operated by an electric motor the disposal unit will grind soft housefold and kitchen waste, vegetable peelings, food scraps, dead flowers and so on, to a slurry that can then be flushed away by running the cold tap.

To take a disposal unit the sink must have a 87.5 mm (3½ in) waste hole instead of the more usual 38 mm (1½ in) outlet. The outlets of stainless steel—but not enamelled—sinks, can be enlarged to the required size with a tool that can usually be obtained from the supplier of the disposal unit.

Waste disposal units are fitted by placing a rubber or plastic washer round the outlet hole and inserting the flange of the unit. The unit is then connected beneath the sink by a snap fastening. The outlet is to a trap in the normal way.

Traps come in all shapes and sizes. Their purpose is to prevent smells – and draughts – from the waste pipe entering the bathroom or kitchen.

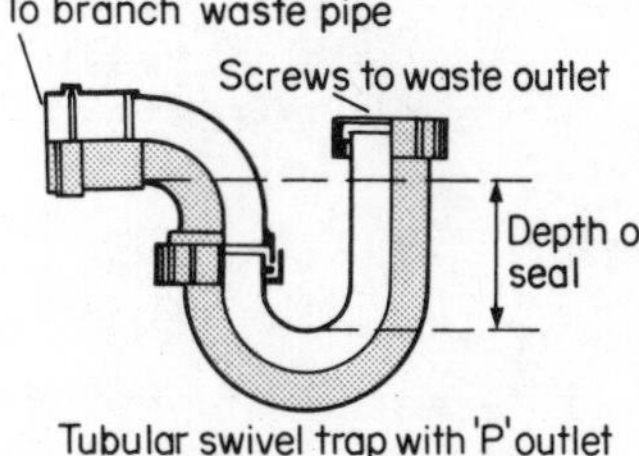

Tubular swivel trap with 'P' outlet

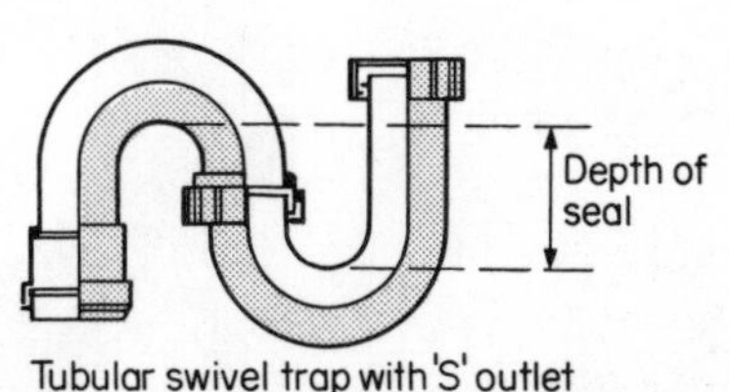

Tubular swivel trap with 'S' outlet

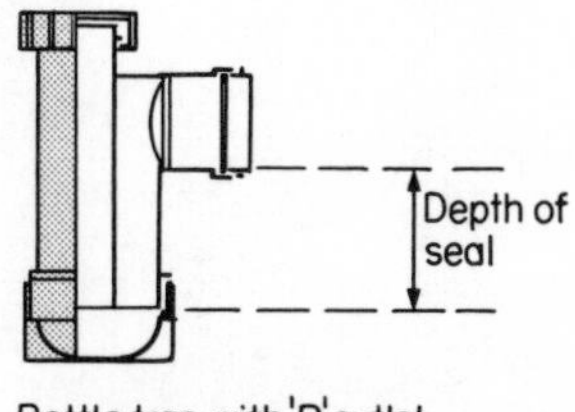

Bottle trap with 'P' outlet

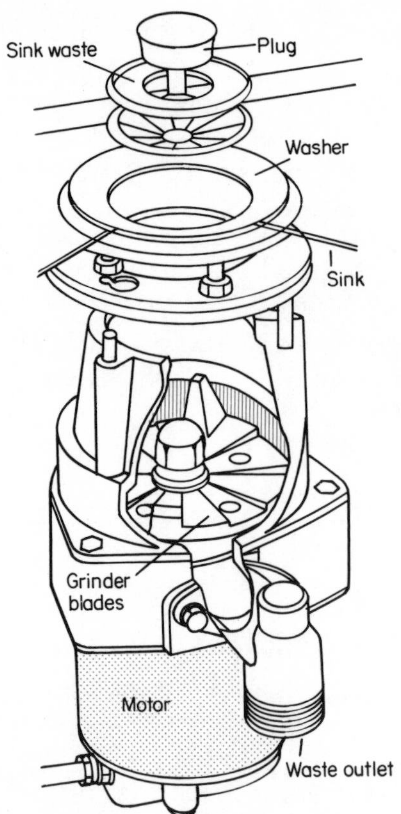

A sink waste disposal unit ('garbage grinder') is easy to fit, provided that your sink has the right size outlet, but an electrician should attend to any new wiring needed.

Replacing a sink

A d.i.y. plumbing job that many householders may wish to undertake is the replacement of an old Belfast sink with a modern sink unit. As with so many jobs of this kind, removing the old appliance is likely to prove far more difficult than fitting the new one!

Disconnect the waste pipe. Lift the old sink off its brackets and remove it. It will be ***very*** heavy. Don't attempt it without assistance.

The cantilever brackets must now be dug out of, or cut flush with, the wall behind the sink.

You will want to fit new pillar taps, or perhaps a sink mixer, into the new sink. Turn off the main stop-cock to cut off the supply to the cold tap and drain the supply pipe to the hot tap as suggested in chapter 4.

Unscrew and remove the old taps and pull forward the water supply pipes. These may be chased into the wall.

Fix the new taps and the sink waste and overflow before moving the unit into position. The new taps should be fitted with a plastic flat washer above the sink and a spacer or top hat washer below. Bed the waste outlet into the hole provided in a non-setting mastic.

Place the sink unit in position. Cut the supply pipes to the correct length and connect them to the tails of the taps or mixer using cap and lining joints.

You ***may*** be able to use the old trap and waste pipe. If not, replace the trap with a plastic or metal bottle trap with an adjustable outlet that will enable you to connect up to the old waste pipe.

Faults in sinks

1. Blocked waste pipe

Sinks, because of the nature of their use, are more likely to develop a choked waste pipe than any other plumbing fitting.

Try plunging with a force cup as suggested in chapter 5 to clear bath wastes. If this proves to be ineffective the trouble is probably due to a solid object jammed in the trap.

All traps have some means of access. Straightforward U traps have screw-in caps at their base. The entire lower part of a bottle trap can be unscrewed and removed.

Place a bucket under the trap before attempting this. The sink will still be full of dirty water! Unscrew the access cap or the base of the bottle trap and probe inside with a piece of flexible wire. The chances are that you will dislodge, amid a flood of water, a hair grip, a match stick or even—if you have young and inquisitive children—a discarded ball-point pen refill!

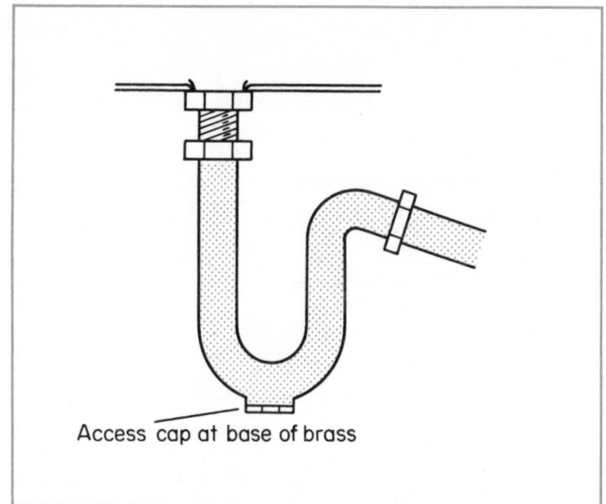

The traditional U-bend trap has an access cap at the base that can be unscrewed to enable you to clear a blockage.

Wash basins

Wash basins—known as 'lavatory basins' or even as 'lavatories' in the trade—like baths, come these days in all shapes, sizes and materials.

The traditional bathroom basin is likely to be made of glazed vitreous china and will be either a pedestal or a wall-hung basin. Pedestal basins conceal the plumbing connections and the pedestal provides additional support. It should not be the ***sole*** support however. Modern pedestal basins are provided with concealed brackets or hangers which are screwed into plugs fixed in the wall behind.

Wall-hung basins are cheaper and are useful where floor space is limited. Before fixing one, make sure that the wall behind is capable of supporting the weight of the basin and, perhaps, an adult leaning upon it. Brick walls are safe enough but internal breeze walls can provide a somewhat dubious support.

Basins have a built-in overflow which connects to an overflow slot in the metal basin waste. When bedding this waste outlet down into the outlet hole make sure that the slot coincides with the outlet of the built-in overflow.

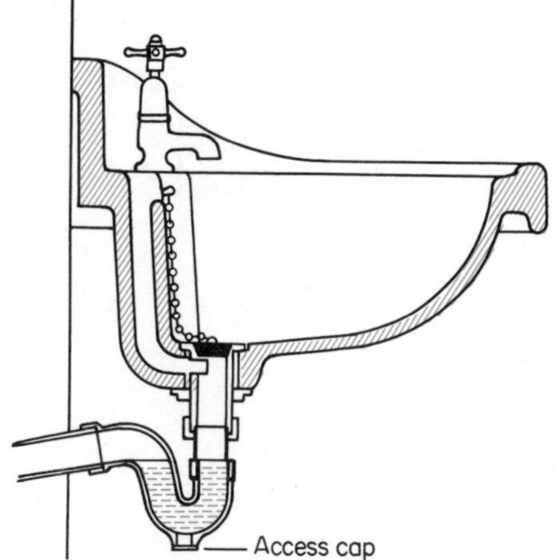

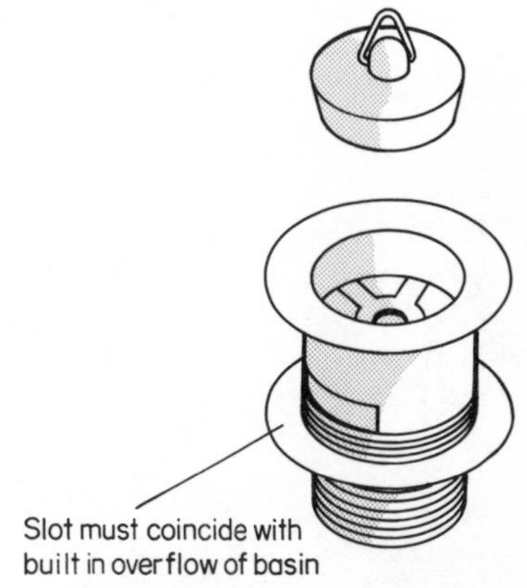

When fitting a ceramic basin of this kind it is essential to make sure that the overflow slot in the basin waste coincides with the built-in overflow of the basin.

When fitting a new basin insert the taps into their holes before placing the basin into position. The taps are fitted in the same way as those of baths and sinks but, with relatively thick ceramic material, you can use a flat, instead of a top-hat, washer underneath. Do not overtighten the back nuts. It is very easy to damage a ceramic basin.

A vanity unit—the modern equivalent of the Edwardian 'wash stand'—is a piece of furniture, perhaps complete with drawers for toiletry, into the surface of which is inset a wash basin. The basin is usually of enamelled steel or of plastic though there are ceramic versions.

Vanity units are most often found in bedrooms where they are the 20th century equivalent of the Edwardian 'wash stand'. They can however also be an attractive and useful piece of furniture.

Occasionally found in very modern bathrooms, the Vanity Unit is particularly useful as a piece of bedroom furniture where it can do a lot towards removing the 8 o'clock queue at the bathroom door.

A vanity unit is fitted in exactly the same way as a sink unit except, of course, that the cold water supply will normally come from a storage cistern and not from the main.

A basin mixer can, in fact, only be fitted if the cold water supply is from a storage cistern. Unlike sink mixers, they do not have separate channels for the hot and cold streams of water. These mix within the body of the fitting and the same rules apply as for a shower mixer.

Some modern basin mixers incorporate 'pop up' waste plugs which eliminate the need for the traditional plug and chain. A control knob, between the hot and cold water controls, can be operated to make the waste plug 'pop up' and allow the basin to empty.

In addition to its use in the bathroom and bedroom the wash basin has a vital role in the lavatory or, as an estate agent would prefer to put it, where there is a 'bathroom and separate toilet'. No civilised person would question the importance of hand washing after visiting this room. This cannot conveniently be done unless there is a basin with hot and cold water installed.

In the past, considerations of space may have made the provision of a basin in the lavatory all but impossible. This is not so today. There are available corner basins, built in basins and basins of every degree of miniaturisation that can be fitted into even the smallest 'smallest room'.

Cold water supply cannot present a problem; it must be available there for the flushing cistern. Where a separate lavatory is remote from the general hot water supply system of the house, a small instantaneous gas or electric water heater can prove invaluable to provide hot water for washing when it is required.

Bidets

Until quite recently the bidet was regarded as an exotic piece of Continental decadence—certainly not a desirable fitting for a respectable British home.

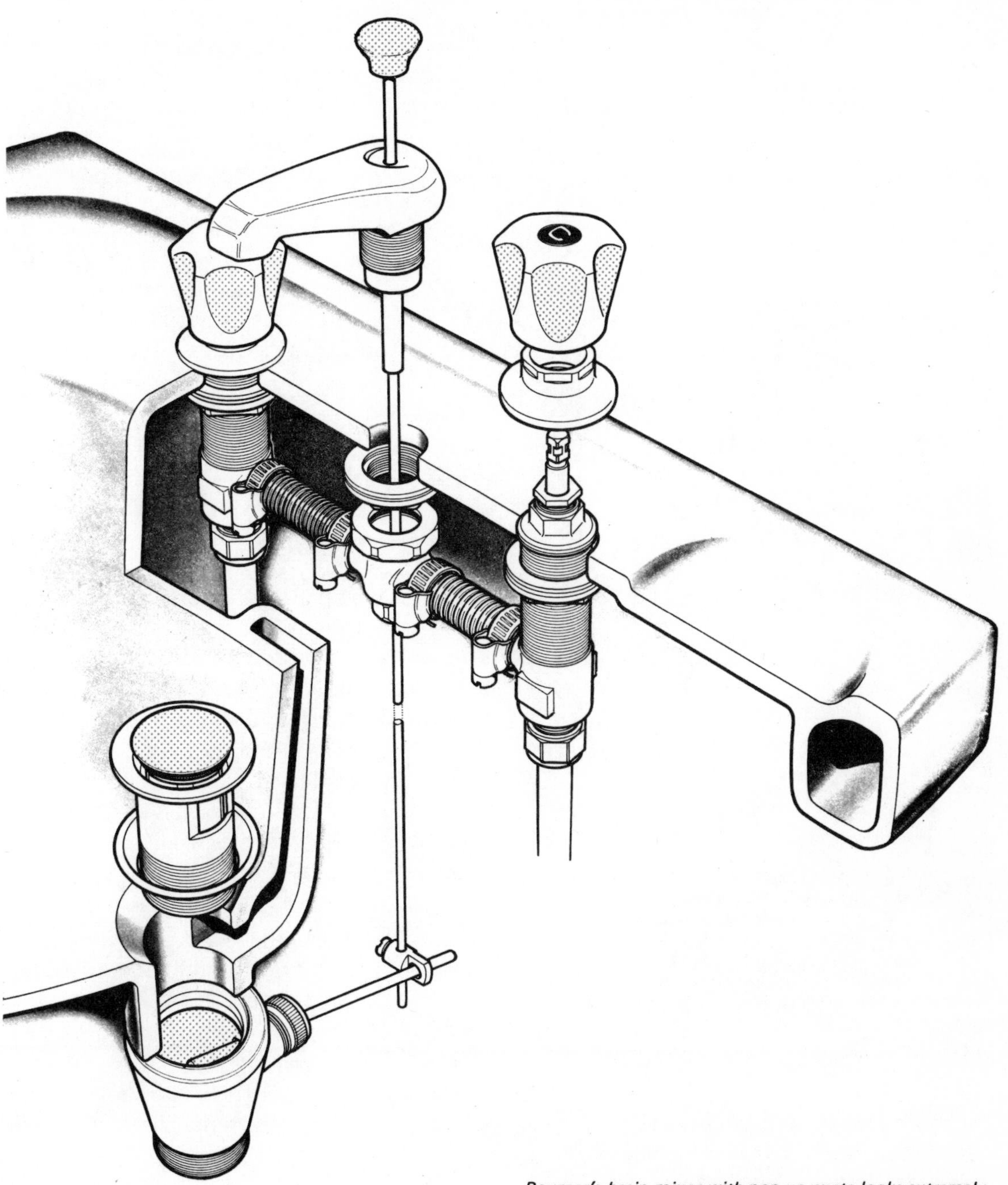

Bourner's basin mixer with pop-up waste looks extremely complicated – until its component parts are revealed. The pop-up waste eliminates the unsightly chain and stopper and permits the basin to be emptied without wetting the hands.

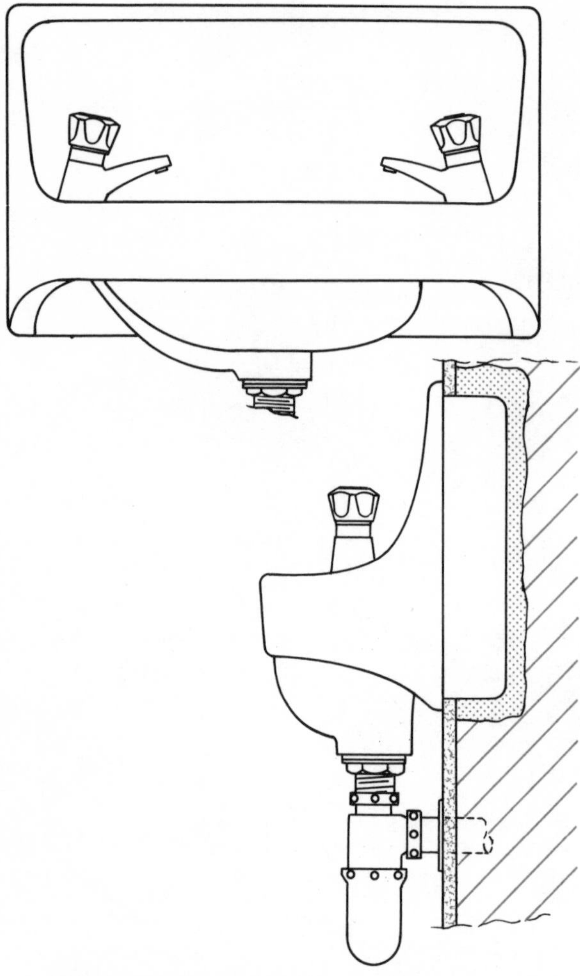

Twyford's built-in wash basin can be fitted into even the smallest lavatory compartment. To fix, cut out brickwork slightly than built-in portion of basin. Line cavity with a weak cement. Place basin in position making allowance for final wall finish. Support basin until cement has set and, finally, apply wall finish. Alternatively a good quality modern adhesive can be used to fix the basin into the wall.

Old habits of thought die hard but these extremely useful pieces of sanitary equipment are gradually gaining acceptance in this country. Certainly, they figure prominently in builders merchants' showrooms and in the glossy catalogues of sanitary ware manufacturers. They are to be found too in growing numbers of British bathrooms.

The bidet is best considered as a specially designed low level wash basin on which the user can sit to wash the lower parts of the body. It can also—as anyone who has spent a walking holiday on the Continent will know—perform a useful secondary function as a foot bath.

There are two kinds of bidet and the differences between them are of considerable importance to the installer.

The simpler kind is, so far as plumbing in is concerned, indistinguishable from a wash basin. Usually described in the catalogue as having an 'over rim water supply', it has two pillar taps, or a mixer, mounted above the fitting at one end. It may have a pop-up or ordinary chain waste stopper and the usual waste outlet and trap.

It can be plumbed into an existing bathroom in exactly the same way as a new wash basin. The hot and cold supply pipes to the bath and basin can be cut, tee junctions inserted, and branch supply pipes taken to the hot and cold taps. It presents no special problems.

The more sophisticated type of bidet is usually referred to as 'rim supply with ascending spray' and special precautions are needed with regard to its installation.

Water enters this kind of bidet in two ways—via a rim, not unlike the flushing rim of a lavatory pan, and via an ascending spray. Warm water passing round the rim warms it and makes it comfortable for use. The ascending spray is directed towards those parts of the body that are to be cleansed.

With any kind of submerged inlet appliance there is a risk of contamination of the water supply by back siphonage.

To avoid this, the hot and cold supplies to the bidet must be taken direct from the hot water storage cylinder and the cold water storage cistern. They must not be taken as branches from pipelines supplying any other fittings. Furthermore the base of the cold water storage cistern must be at least 2.75 m (9 ft) above the inlet to the bidet.

From the point of view of drainage, bidets are regarded as being 'waste', not 'soil' fittings. Where the two-pipe system of above ground drainage is provided (see next chapter) the waste pipe from a bidet should discharge over a gully in the same way as that of a bath, basin or sink. It should not, like the outlet of a lavatory pan, be connected directly to the drain or soil pipe.

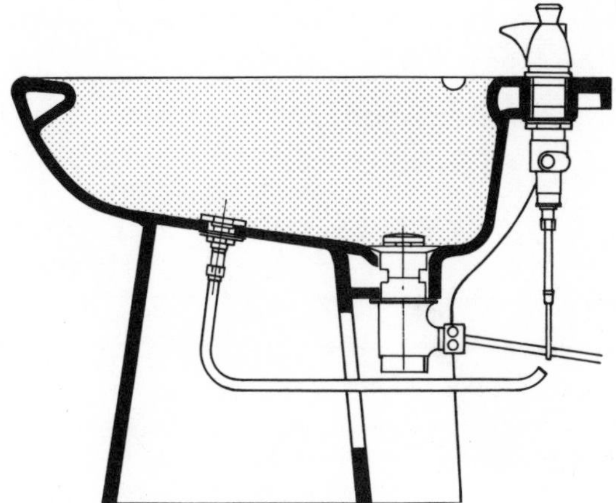

A rim supply, or ascending spray, bidet has a flushing rim round which water passes to fill the bidet and warm the seat for the user and an ascending spray directed towards the parts of the body to be cleansed. To prevent contamination of water supplies special design precautions are necessary in the installation of this kind of bidet.

An over-rim supply bidet is essentially a low level wash basin and presents no special plumbing problems.

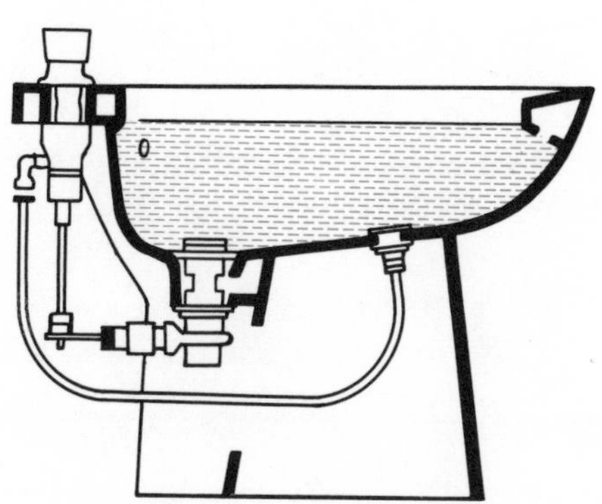

Rim supply with ascending spray

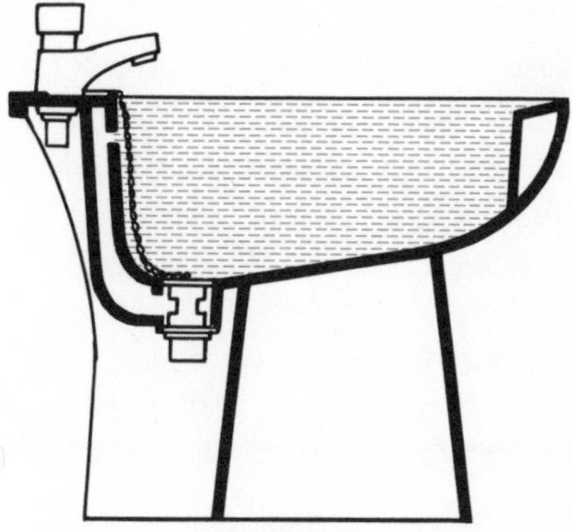

Over-rim supply bidet

Chapter 8
The drains

There have been tremendous changes both in design and materials in domestic drainage during the past two decades. These changes have been particularly revolutionary in the part of the drainage system that is above ground level.

If your home was built before the early 1960s it will almost certainly have a two-pipe drainage system. This system drew a distinction between 'soil fittings' (lavatory suites and urinals) and 'waste fittings' (sinks, baths, basins and bidets). The outlets from soil fittings were connected directly to the drainage system. The outlets from waste fittings were disconnected from the drainage system and entered it only via an external yard gulley.

A heavy cast iron or asbestos cement soil pipe would rise, to above eaves level, against an external wall of the house. This would be open ended to provide a high level ventilator to the drain. Into it would be connected the branch soil-pipe from any upstairs lavatory suite. The outlets of ground floor lavatories were connected directly to the underground drainage system by a branch drain joining the main drain at the nearest inspection chamber or drain manhole.

The outlets from ground floor baths, sinks or basins protruded through the wall of the house to discharge over a yard gully which would, in its turn, be connected to the main drainage system via a branch drain.

Upstairs waste fittings (baths, basins and bidets) presented special problems. In the provinces it was usual for them to discharge in the open air, over a rain water hopper head. This would take the wastes to ground level through a length of waste or rain-water pipe to discharge, like the sink waste, over a yard gully.

Hopper heads were, and are, smelly, insanitary fittings. Soapy water discharged from bath and basin waste pipes dries and decomposes on their internal surfaces, producing unpleasant smells in the immediate vicinity of bedroom windows. Draughts, whistling up the open-ended main waste pipe, discharge smells from the gully in the same vicinity.

In some towns, notably London, the insanitary nature of these fittings was recognised and they were banned by local drainage bye-laws. In these areas the main waste pipe, like the soil pipe, had to extend to above eaves level. Branch waste pipes were connected into it but, as with the hopper-head arrangement, the bottom end of the waste pipe still discharged into the open air above a yard gully.

The Building Regulations of the mid-1960s, by requiring that all soil and waste pipes should be contained within the fabric of the building, hastened the almost universal adoption of the 'single stack' system for all above-ground drainage work. In the single stack system the distinction between 'soil' and 'waste' fittings is discarded and waste outlets from all sanitary appliances discharge into one main soil and waste stack, connected directly to the underground drain. Apart from the open end of this stack, protruding through the roof, the whole of the drainage is contained within the walls of the building.

Take a look at a house built in the last few years and note the difference from one built fifteen years or more ago. Gone is the hopper head, the waste pipe and the soil pipe climbing, as someone once said 'like petrified worms', up the wall of the house. Apart from the rain water gutters and down-pipes the only evidence of a drainage

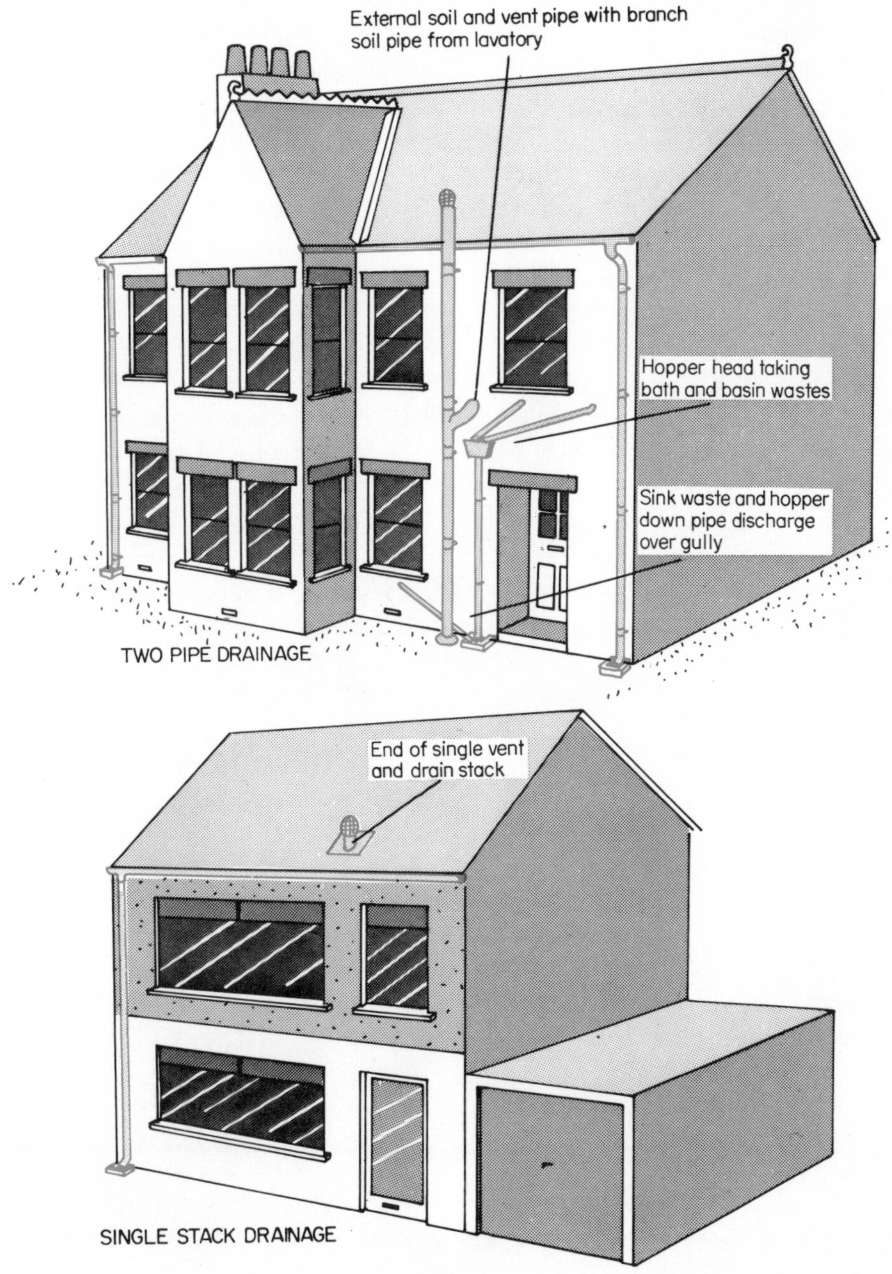

The elevation of a house built before the mid-'60s will probably look like the top diagram. A branch soil pipe from an upstairs w.c. connects to a main external soil and vent pipe rising, open ended, to above eaves level. A hopper head collects wastes from an upstairs bath and basin and discharges them over the grid of a yard gully. A ground floor sink waste discharges over the same gully.

Apart from rain water guttering and down pipes, all that is likely to be visible of the plumbing of a modern house (shown in the lower diagram) is a few inches of PVC vent and drain stack protruding from the roof. The main stack, within the fabric of the building, may take w.c., bath, basin and sink wastes.

system that you will see will be a few inches of capped PVC tubing protruding from the roof above the bathroom.

When first introduced into this country, from the other side of the Atlantic, the single stack drainage system was regarded with suspicion by plumbers and public health engineers alike.

I well recall—at Battersea Polytechnic in 1947—earning the lecturer's unstinted praise for an essay in which I unequivocally condemned the single stack drainage system. It would, I said, undo all the progress made in sanitary science during the previous half century and would inevitably result in drain smells pervading domestic bathrooms and kitchens.

I pointed out the dangers of loss of seal in sink and basin traps from self-siphonage and induced siphonage. Basin traps had always had a tendency to self-siphonage. They were shallow (often with only a 1 in or 1½ in seal) and the discharge of the basin into a small diameter branch waste pipe often produced the partial vacuum that is a prerequisite of siphonic action. With the two-pipe system temporary loss of seal was of no great importance. Only smells from the short branch waste could enter the room. With a single stack system such loss of seal would—so I said—have catastrophic results.

Then again, siphonage of basin traps would be induced by bath wastes discharging past the connection of the basin waste to the main stack and aspirating the air from between the trap and the waste connection.

I pointed out too, that any blockage at the foot of the main soil and waste stack would make itself known by sewage flowing into the ground floor kitchen sink. If the single stack drainage system served a block of flats, residents on the upper floors would carry on using the drainage system oblivious of the havoc that they were creating in the flat of their ground floor neighbour.

The points that I made in that essay, that earned me full marks in 1947 (and would earn a student none today!), were valid enough. The introduction of the single stack system could have produced all the disasters that I so confidently predicted.

The dangers of siphonage, compression and blockage are real ones but they can be—and have been—overcome by careful planning and design.

Design of the waste system

For really successful single stack drainage the house needs to be designed round the plumbing system. Waste pipes from all appliances should be kept as short as possible. They should be laid with minimal falls. All fittings should be provided with 75 mm (3 in) deep seal traps. The connections to the stack pipe should be carefully planned to prevent any risk of the discharges from the lavatory suite outlet fouling and possibly blocking the outlets from other appliances.

The connection between the vertical waste stack and the underground drain must be made with an easy bend and the lowest connection to the stack pipe should be at least 760 mm (2 ft 6 in) above the level of the underground drain.

The design of the waste from the wash basin is particularly critical. These waste pipes are of small diameter (1¼ in or 30 mm) and are likely to be filled with water when the basin is emptied. There is therefore a considerable risk of the water in the trap being siphoned out.

To prevent this the basin waste must be laid with a very slight fall and should be no more than 1.68 m (5 ft 6 in) in length. Where circumstances compel a longer basin waste

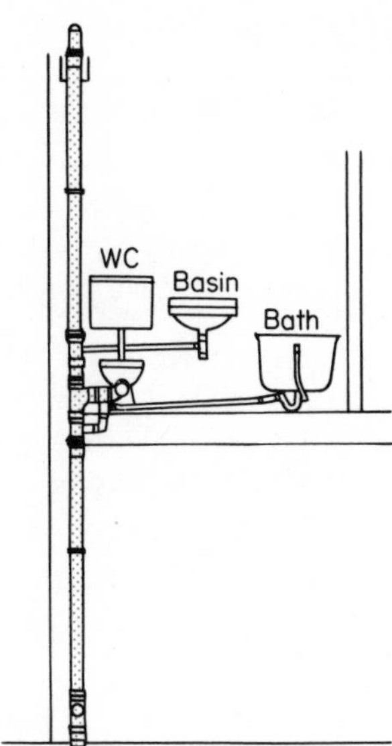

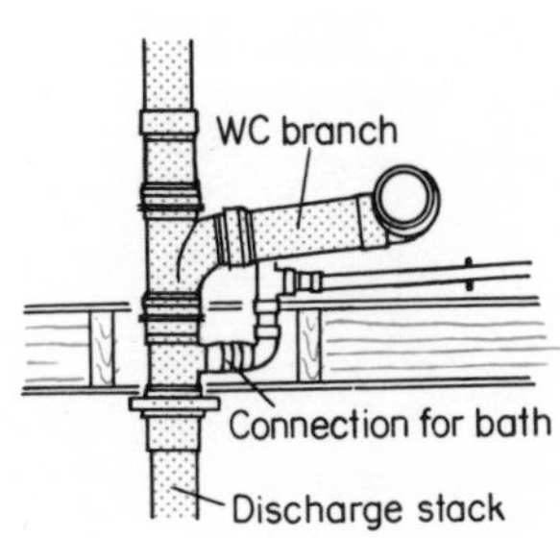

In order to avoid the danger of the outlet from the bath becoming fouled with discharges from the w.c. the bath waste – in a conventional single stack system – must be offset and taken into the main drainage stack at a point below the w.c. connection.

pipe, a larger diameter pipe may be used or, alternatively, a small ventilating pipe may be taken from behind the basin trap to connect to the main stack at least 1 m (3 ft) above the highest waste connection.

Prevention of the fouling of waste pipe outlets by lavatory suite discharges presents problems so far as the bath waste is concerned. One way to get over the difficulty is to offset the pipework through the floor so that it enters the main stack in the ceiling space or below it.

This is by no means a convenient arrangement and an alternative solution is to use the Marley collar boss which permits the bath waste to enter the stack immediately below the lavatory suite waste via a collar that protects it from any possibility of fouling or blockage.

Polyvinyl chloride (PVC or vinyl for short) is used almost universally for domestic single-stack above ground drainage. It may be joined by solvent welding or by ring-seal joints (see Chapter 11). Because this material expands as hot water passes through it, special expansion joints should be fitted where any straight run of waste pipe exceeds 1.8 m (6 ft) in length.

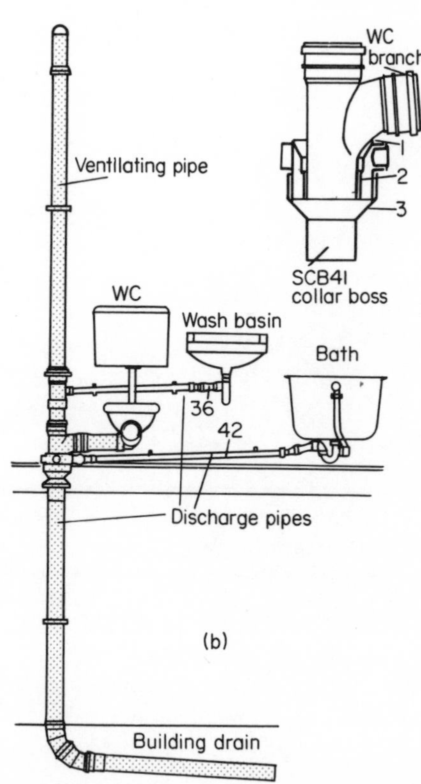

The Marley collar boss avoids the necessity of offsetting the bath waste pipe while affording it absolute protection from w.c. discharge.

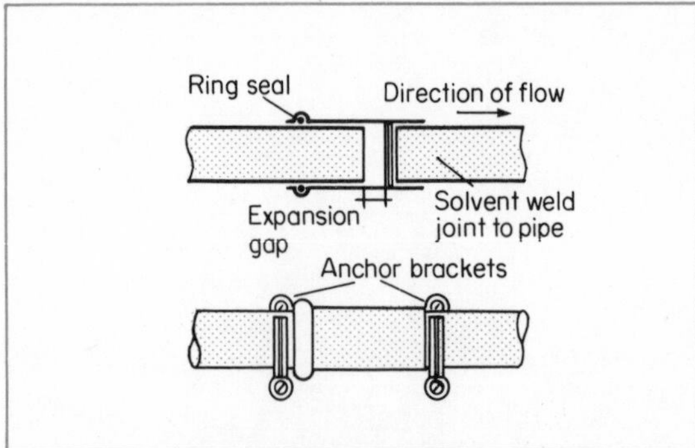

PVC tubing expands as hot water runs through it. The introduction of an expansion joint into a long run of tubing accommodates this expansion.

Although the waste pipe from a ground floor sink may be, and frequently is, connected to the main drainage stack, there is something to be said for taking it to an external gulley in the traditional way. The risk of a blockage at the base of a properly designed stack system may be remote. It is not, however, altogether impossible and the thought of sewage backing up into the kitchen sink is one that lacks appeal.

Furthermore, if the sink waste pipe discharges over a gully, it may be possible to lay the underground drain at a shallower depth—and depth means money where drainage is concerned!

If the sink waste pipe is taken to a gully it should enter it above water level but below the grid. This makes sure that the full force of the sink discharge is available to flush and cleanse the gully. It also ensures that autumn leaves blowing on to the grid cannot result in a flooded yard.

Underground drains

So far as underground drainage is concerned the changes that have taken place in recent years have related to materials rather than to design. Underground drains must still be laid in straight lines at a regular self-cleansing gradient. There must still be means of access for rodding every part of the drain. Branch drains must still connect to the main drain, in the direction of flow, at inspection chambers or other access points.

If your home was built more than about twenty-five years ago the underground drains will consist of salt-glazed stoneware drain pipes, 2 ft long and connected with cement joints. They will be laid on a 6 in deep concrete bed and concrete will be haunched over them to protect them and to give the entire drainage structure some stability to reduce the risk of damage in the event of ground settlement.

They will, almost certainly, be laid to a fall of 1 in 40 (3 in in 10 ft). At each change of direction and wherever branch drains connect to the main drain there will be a brick built inspection chamber or manhole with an iron cover.

The drain passes through the inspection chamber in a half-channel with concrete benching on either side to prevent the walls of the chamber being fouled. The brick walls of the chamber *may* be rendered with cement and sand to make them watertight. This is now recognised as bad practice. The rendering is all too liable to flake off and block the drain. If concrete rendering is necessary it should be on the external walls of the chamber, between the walls and the surrounding earth.

The inspection chamber

Just inside your front gate will probably be a final inspection chamber containing an 'intercepting' or 'disconnecting' trap. The object of this trap, almost universally discarded in modern drainage systems, was to prevent gases, and perhaps rats, from the sewer entering the house drains.

The intercepting trap, where it exists, is undoubtedly the commonest site of drain blockage. Its disadvantages are very real and its benefits largely illusory. There should not *be* rats in a modern well-maintained

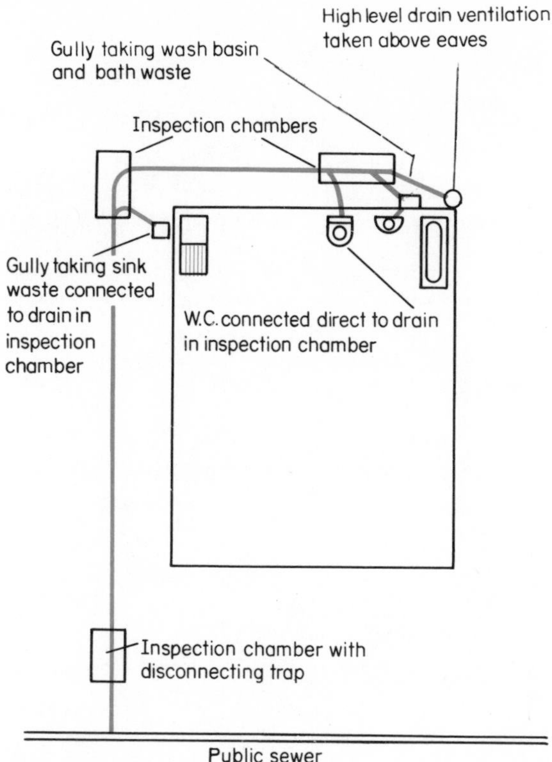

A plan of a traditional drainage system shows bath, basin and sink wastes disconnected from the drain by means of gullies. Inspection chambers are provided at junctions and changes of direction and an inspection chamber with an intercepting trap disconnects the house drainage from the public sewer.

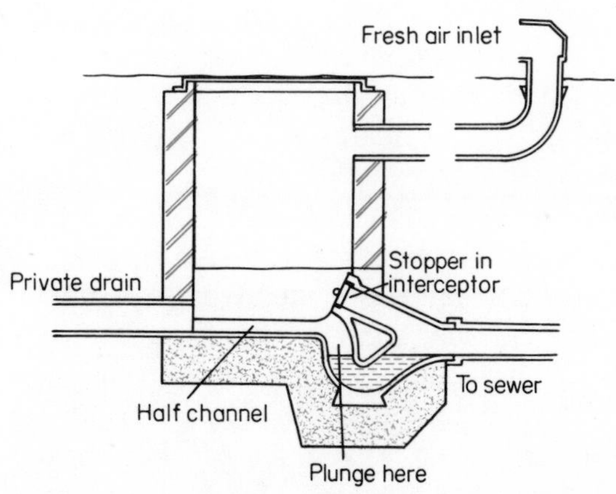

Intercepting or disconnecting traps are now regarded as absolute. Their purpose was to prevent sewer gases or rats entering the house drains. They were a frequent site of blockage and the fresh air inlets associated with them were a common source of 'drain smells'.

sewer and, if the sewer is ventilated through the soil and vent stacks of every house connected to it, the sewer should be free of offensive gases.

The intercepting trap incorporates a rodding arm, closed by a stoneware stopper, through which it is possible to rod the drain right through to the sewer. It should perhaps be emphasised at this point that the householder's responsibility for his drain does not end at the boundary of his property. He is responsible for any blockage or defect that may occur in the drain right up to the point at which it connects to the public sewer.

If your final inspection chamber has an intercepting trap it will also probably have a 'fresh air inlet'. This is a metal box with a grille at the front against which is hinged a mica flap. You will find this near the trap.

The idea was that air could enter the drain at this point, flush through it, and escape via the soil and vent pipe. It rarely worked quite like that. Fresh air inlets are particularly susceptible to accidental damage. A stroll down any suburban street developed between the '30s and '50s will reveal a dozen fresh air inlets with their metal boxes broken or their mica flaps jammed. Many will have been simply sealed off by their owners as being a source of smells and performing no useful purpose.

Layout of underground drains

Modern underground drains are likely to be of PVC or pitch fibre with simple push-on ring seal joints. They are not laid on a concrete base though preparation of the bed on which they lie remains important. Where the sub-soil consists of heavy clay or chalk it may be necessary to prepare an imported bed of gravel to form a base.

Gradients are a good deal less than they were with the many-jointed stoneware drains. Too steep a drain gradient is as likely to create a blockage as a too shallow one; the liquid will run on, leaving solid matter deposited in the pipe. Falls of 1 in 60 to 1 in 70 are quite usual.

There will still be inspection chambers at junctions and changes of direction. These may well be built of brickwork and have stoneware half-channels. They may, on the other hand, be prefabricated in fibre-glass reinforced plastic or constructed on site from concrete sections. Marley Extrusions have produced a sealed drain access system which bears a close resemblance to the sealed iron drainage systems sometimes provided in very high class building work in the pre-war years.

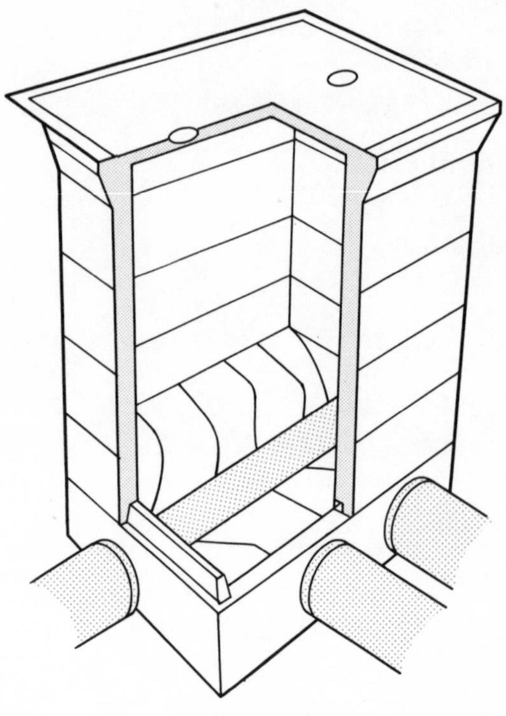

Sectional concrete inspection chambers can be quickly installed by an inexperienced worker.

It must be emphasised that although, in a sense, drainage work is 'easier' than it was with the old methods and materials, there is much less margin for error in design and installation; particularly where above ground work is concerned. Taking a waste pipe from an additional wash basin or a newly installed shower cabinet to an existing gully or hopper head could not possibly have any adverse effect upon a traditional two-pipe drainage system. Making such a connection wrongly into a single stack system could have very serious results.

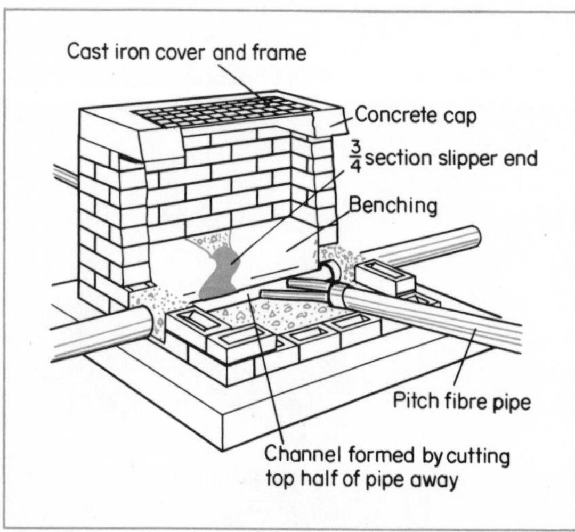

A conventional brick-built inspection chamber with a pitch fibre drain running through it. Note the concrete benching beside the half channel in the middle of the chamber and the fact that branches join the main drain 'in the direction of the flow'.

All new drainage work—even that involved in the provision of an extra wash basin—must comply with the Building Regulations that are enforced by your local District or Borough Council.

Before embarking upon any project of this kind and, above all, before committing yourself to any expense, have an informal chat with the Council officer responsible for enforcing the Building Regulations relating to drainage. It could be the Environmental Health Officer or the Building Control Officer that you will need to see. He will tell you what the Council's requirements will be and may well be prepared to give you some useful on-the-spot advice.

Do not believe all that you may read in the national press about 'Town Hall bureaucrats'. They are there to help—not to try to 'catch you out'!

Gutters and down-pipes

Collection and disposal of rain water falling on the roof is an important aspect of domestic drainage. You will find that the material of which the gutters and down-pipes are made give a pretty good indication of the age of a house.

Pre-war, and immediately post-war, houses always had cast iron gutters and down pipes. These are strong, hard wearing and give plenty of support to ladders propped against them.

Their only real disadvantage, and it is a major one, is their liability to corrode unless protected by paint. As well as painting externally they must be cleaned out and treated internally with bituminous paint at regular intervals. This is a vital chore that adds time, and money, to house maintenance.

Shortly after the war there was a brief vogue for asbestos cement gutters and down pipes. These do not rust and need no decoration. They are rather thick, heavy and clumsy in appearance though and can break all too easily if a ladder is dropped against them. Similarly, the downpipes can easily shatter on accidental contact with, for instance, a wheel barrow or a lawn mower.

Most newly-built houses nowadays have PVC gutters and downpipes and rainwater goods of this material are increasingly used for replacement work. PVC is lightweight, easily fitted—for a replacement job you're likely to find removing the old gutters far more difficult than fitting the new ones—attractive in appearance, cannot corrode and needs no decoration.

It is not however strong enough to support a ladder. If you need to get to the roof, rest your ladder against either the wall or the fascia board. The Marley PVC rain water system is illustrated on the next page.

How many down pipes? For most houses one at the back and one at the front of the house is ample. An arrangement of this kind will be adequate for a total roof area of 2800—1400 sq ft front and 1400 sq ft back.

Fall should be minimal. For domestic roof drainage a fall of 1 in in 50 ft is sufficient. If the fall is noticeable from ground level it will spoil the appearance of the house.

The ultimate destination of the rainwater will depend upon the policy of the local sewerage authority (prior to 1 April 1974 the District or Borough Council, since that date the local Water Authority). In some areas rainwater gullies are allowed to be connected directly to the household's main drain.

Other areas, in order to reduce the cost of sewage treatment and to guard against surcharging of the foul sewer during during periods of heavy rain, either provide a separate surface water sewerage system or require the provision of a soakaway for the disposal of water draining from roofs.

Faults with drainage systems

1. Blockages

Blockages in traps and waste pipes have already been dealt with. A blockage in the underground drain may make itself known by water flooding from a yard gully or from under the rim of a drain inspection cover. Another common first symptom is for a lavatory suite, when flushed, to fill almost to the rim of the pan and for the water then—very slowly—to subside.

If flooding from a yard gully is the first

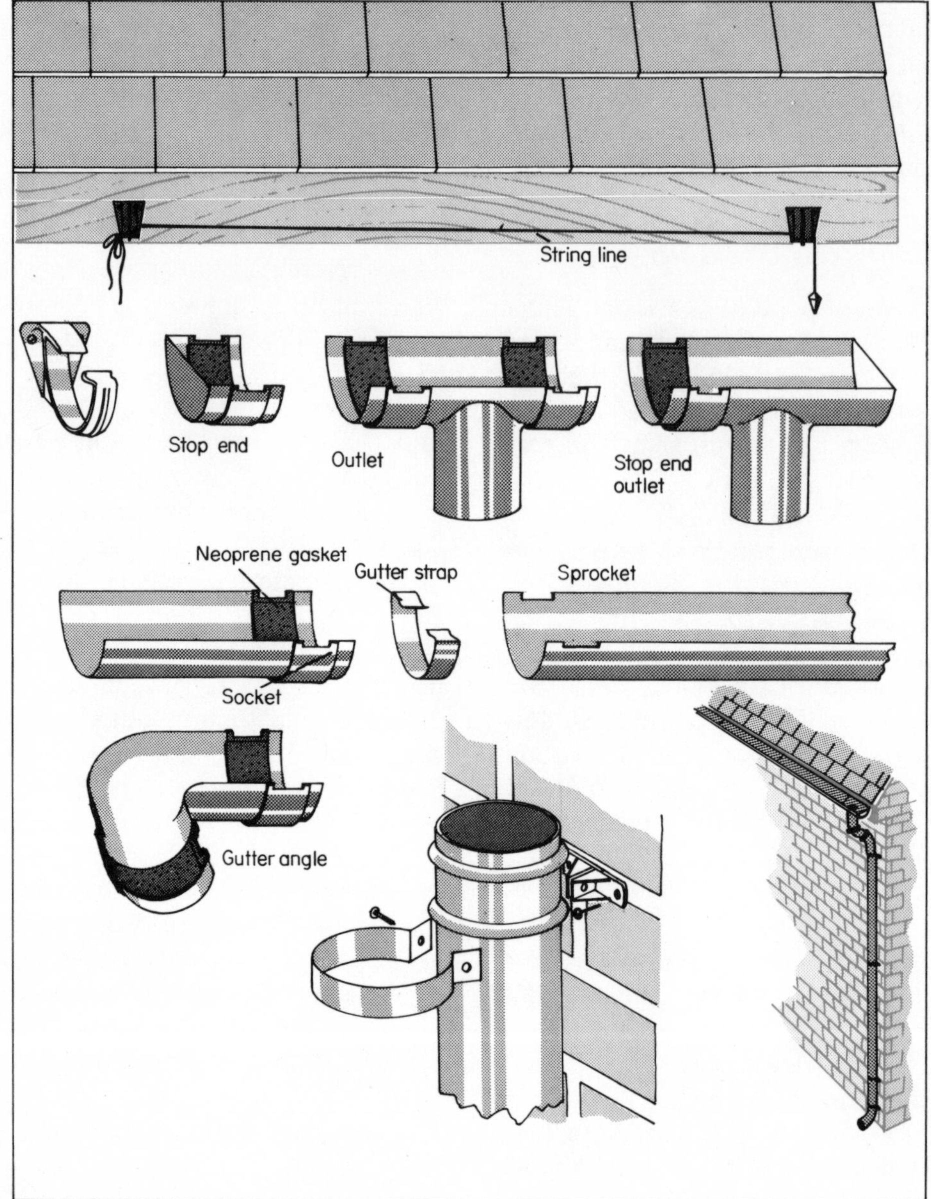

PVC rain water gutters and down pipes are light, easily fixed and need no decoration. The Marley system is illustrated.

sign, check first of all that the trouble isn't due to the grid being choked with leaves or similar debris.

Next, raise the covers of the drain inspection chambers, beginning with the one furthest from the sewer. If this chamber is flooded but the next one is empty then the blockage must obviously lie between the two chambers.

You will need a set of drain rods or sweeps rods to clear it. Screw two or three rods together and lower the end into the flooded chamber. Feel for the half-channel at the base and push into the drain towards the blockage. Screw more rods on to the end and thrust along the drain until the obstruction is reached and dislodged.

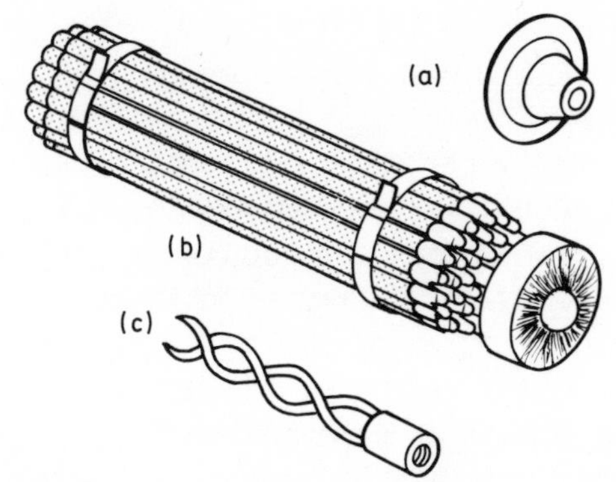

A useful kit for drain clearance comprises a set of drain rods with a 4 in drain plunger (a), a drain cleaning brush (b), and a screw for removing difficult obstructions (c)

If your drain has an intercepting trap the chances are that it is here that the blockage will be located. To clear it you will need a drain plunger—a 4 in diameter rubber disc—screwed on to the end of a couple of drain rods.

Lower the plunger into the inspection chamber. Feel for the half channel and push the plunger along until you can feel the drop into the trap. Plunge down sharply two or three times and, almost certainly, there will be a gurgle and the water level in the inspection chambers will fall as the sewage runs away.

A blockage between the intercepting trap and the sewer is, fortunately, relatively rare. To clear it you need to knock the stopper out of the intercepting trap rodding arm and rod through to the sewer.

When using drain rods there is one important point to remember. Twisting them clockwise will help them along the drain and will also help you to withdraw them when the blockage has been cleared. ***Never twist anti-clockwise.*** If you do you will unscrew the rods and leave some of them in the drain.

2. Drain smells

If the smell is in the house check on nearby yard gullies or rain water hopper heads. They may take only soapy water from baths and basins—but this can have an objectionable smell as it dries and decomposes. Cleanse with hot soda water.

Consider the possibility that a trap—of a wash basin perhaps—connected to a single stack system, may have siphoned out. You should be able to see the surface of the water seal below the grid of the waste. If this is the cause of the trouble you may have to consider ventilating the waste pipe to prevent self siphonage. Seek the advice of the Council's Environmental Health Officer.

Remember too that the overflow of a wash basin can be very difficult to clean and may smell from a build-up of soapy water.

A smell of drains in the garden is usually an indication of a choked, or partially choked, drain.

A very common form of partial blockage occurs when the stopper falls out of the rodding arm of an interceptor trap. This may occur as a result of pressure building up within the sewer during a heavy rainstorm.

The stopper falls into the intercepting trap and produces a blockage—but it is not a blockage that becomes immediately apparent. Water level rises in the intercepting trap inspection chamber but the sewage can escape by flowing down the now open rodding arm.

This may take place over a period of several weeks during which time the lower part of the intercepting chamber becomes a minature cesspool, the sewage in it becomes fouler and fouler as a result of decomposition.

Eventually the blockage makes itself all too apparent by the offensive smell that welcomes visitors near the front gate!

When this trouble occurs it is a good idea to remove the stopper altogether. Replace it in the inlet to the rodding arm with a disc of glass or slate, cut to size and lightly cemented into position. On the rare occasions that it is necessary to rod through to the sewer this disc can be broken with a crowbar and afterwards replaced.

Don't overlook the possibility that an offensive smell may ***not*** emanate from the drains or the sewer. A leaking gas main could be a cause. Dead bushes or shrubs in a hedge or shrubbery under which the main passes can give an indication of this.

There used to be manufactured a particular kind of plastic used for electric fittings which exuded an exceptionally unpleasant 'fishy' smell when it became hot. I discovered this the hard way after spending hours trying to find which tenant in a block of flats always 'cooked up cat food' just at dusk—after the electric lights had been switched on!

These fittings are happily obsolete but I would be surprised if there are not some still in use.

3. Leaky drains

If a persistent patch of dampness on a path or on a basement wall leads you to suspect that the drains may be leaking, always seek the advice of the Council's Environmental Health Officer. He can check this by smoke testing or colour testing.

4. Water overflowing from rain water gutters or leaking from joints in downpipes

Check on the fall of the gutters and check that they are unobstructed. An astonishing amount of silt—not to mention children's balls and similar objects—can accumulate in rain water gutters during the course of a year.

A flow of water, during heavy rain, from joints in downpipes could indicate a blocked rain water drain. It is more likely though, to be a sign of a silted up soakaway.

Soakaways usually consist of a pit, about 5 ft deep and 4 ft square in plan, filled to within about 1 ft of the surface with brick rubble and with the surface soil made good on top.

After a period of use the spaces between the rubble become clogged with silt and the soakaway needs to be dug out and remade. There are concrete sectional soakaways on the market, with holes in the side through which water can soak into the surrounding soil. These have manhole covers to give access for digging out when necessary.

It should be added that soakaways are unlikely to be very successful where the soil is heavy and the level of subsoil water is high. In prolonged periods of heavy rain any soakaway may prove to be incapable of coping with the volume of water flowing into it.

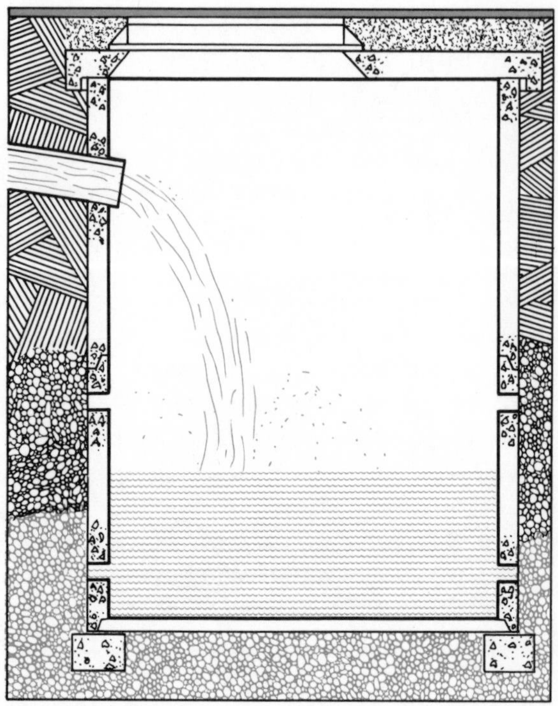

A precast concrete soakaway can easily be opened up and dug out when it becomes choked with silt. Like any soakaway it will be effective only where the subsoil is light and absorbent.

Chapter 9
Hard water problems

'Pure water'—the hydrogen oxide (H_2O), with which every schoolboy is familiar—is unknown in nature. The reason for this is that water is ***the*** great solvent. There are few gases or solids that it cannot take into solution to some extent.

Rainwater is distilled by nature, it leaves the clouds free of impurities. Yet in its brief passage from raincloud to earth it dissolves measurable amounts of carbon dioxide and sulphur dioxide gas. Despite smoke-free zones it may still, over industrial towns, bring down with it several tons of soot, grit and dust per square mile every year.

The water that we draw from our taps fell originally as rain. Yet before it reached the water authority's reservoir it had flowed in a river or stream, or had soaked into the ground through layers of rock, to be pumped up again from deep, natural underground reservoirs. All the surfaces and gases with which it came into contact added their contribution and can be found in solution.

In the reservoirs of the water authority, more chemicals may be added. Some are to destroy the germs of disease with which the water may have become contaminated. Other chemicals, in certain areas, are added to reduce the risk of dental decay.

Some chemicals produce the condition in water known as hardness. Hard water is wasteful with soap, turning it into an insoluble curd. It can ruin woollens washed in it by matting the wool with undissolved soap. It can make a misery of hair washing. It furs up kettles and hot water systems.

It is wasteful too with fuel. A mere 1/8 in of scale inside a gas, oil fired or solid fuel boiler can raise the cost of using the appliance by as much as one third. I have already mentioned, in Chapter 2, how scale can result in boiler damage as a result of the scale insulating the metal of the boiler from the cooling effect of the circulating water.

Hardness is the result of water, in its journey from rain-cloud to reservoir, taking into solution the bicarbonates and sulphates of calcium and magnesium. Water from deep wells and from rivers with a chalky bed is most likely to be hard. This applies to most public supplies in the southern and eastern areas of Britain. Water from mountain catchment areas and from upland reservoirs—the holiday areas of the north and west—is likely to be soft.

Kettle fur and boiler scale

When hard water is heated to temperatures above about 140° F (60° C), carbon dioxide is given off and the dissolved bicarbonates of calcium and magnesium are changed into insoluble carbonates. These are deposited to form kettle fur or boiler scale.

This kind of hardness, which can be removed by boiling, is called temporary hardness. The hardness caused by the sulphates of calcium and magnesium cannot be removed in this way. This is called permanent hardness.

A chemical analysis of a public water supply is likely to quote a figure for temporary hardness, permanent hardness and, the sum of the two, total hardness.

Means of preventing the formation of boiler scale - temperature control, the use of an indirect hot water system, the use of chemical scale inhibitors - have been discussed in Chapters 2 and 3.

A more radical solution to this, and to all other hard water problems, is to install a modern base exchange or ion exchange mains water softener. This will ensure that every drop of water that flows into your home is as soft as—or even softer than—the water that you enjoyed during your last holiday in Cornwall or the Western Highlands.

Water softeners

The principle on which these water softeners operate was discovered from the observation that when hard water flowed through natural zeolite sand, the chemicals causing hardness, 'exchanged bases' with chemicals in the sand. Sodium zeolite would, for instance, become calcium zeolite and the calcium bicarbonate in the water would be changed into sodium bicarbonate which does not cause hardness.

Modern water softeners use a synthetic resin instead of a natural zeolite sand, but the way in which they operate remains the same.

Typically, a mains water softener consists of a cylinder made either of plastic, reinforced with glass fibre, or of metal protected against corrosion, connected to the water main and with a soft water outlet. Hard water flows in from the main and is softened as it passes through the resin bed.

Eventually the resin bed needs regeneration if it is to continue the softening process. This is done quite simply by running through it a strong solution of sodium chloride (common salt).

When the base exchange, or ion exchange, material in a water softener becomes exhausted it can be regenerated. First, a back wash to loosen the material and to remove any grit or debris that has collected on its surface. Next, salt is added and flushed through the softening material. This regenerates the softener which can then be brought back into use.

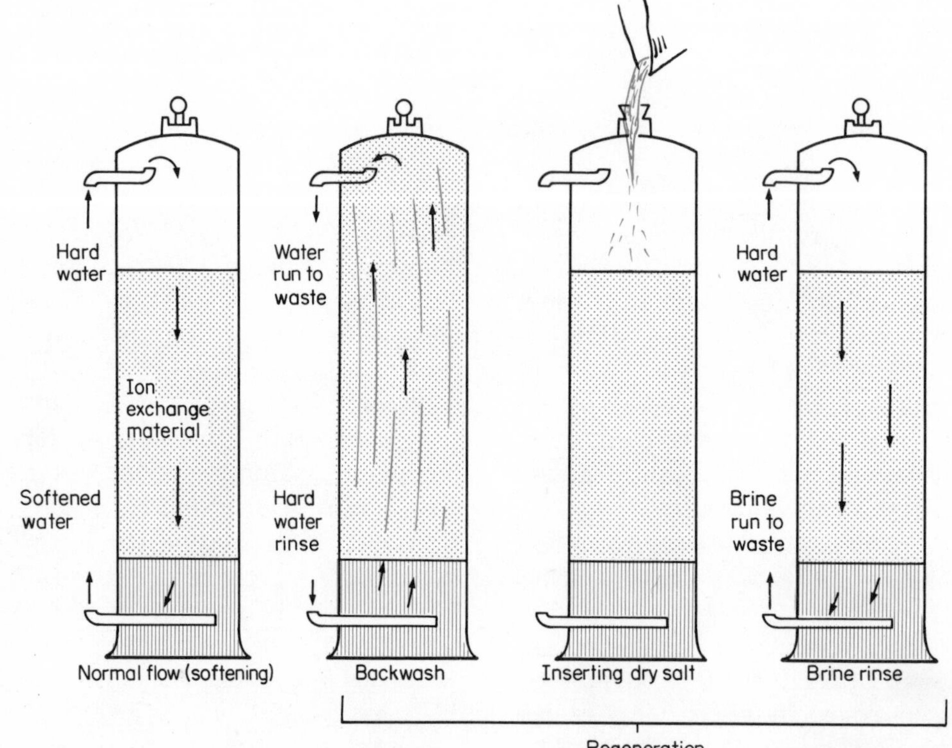

Recharging water softeners

Recharging the water softener with salt, usually weekly or fortnightly, used to be a somewhat tedious chore. The filter had to be back-washed to remove any debris that the resin had filtered out. Salt was then poured into a salt-space at the top of the appliance and the valves of the softener opened to allow water to flush the salt through the resin bed and then flow to waste. When all taste of salt had gone, the waste valve was closed and the normal softened water supply to the house resumed.

Nowadays, automatic controls make life easier for the water softener owner. A modern water softener is likely to incorporate a salt reservoir capable of containing sufficient salt for anything between 15 and 30 or more regenerations. An electric time control prompts the softener to regenerate itself automatically at a preset interval. The householder has only to recharge the salt reservoir—possibly no more often than once every eight or nine months.

Those who live in a rural area and have their own septic tank sewage treatment plant should make sure that the salty flush from regeneration does not flow into the household drains. Salt is a disinfectant which does not distinguish between benevolent and harmful bacteria. Septic tank sewage treatment is a bacterial process. Regular flushing with a strong salt solution could seriously affect its action.

For those who cannot afford a mains water softener there are less expensive ways of obtaining limited quantities of softened water.

Most manufacturers of water softeners, make small portable softeners as well as their big mains models.

A typical portable softener, the Permutit Water Midge, is only 18 in high with a maximum diameter of 7 in. It has a flexible hose suitable for connections to ½ in or ¾ in taps and, with a water supply of average (18°) hardness, will soften about 150 gallons of water between regenerations. Regeneration is carried out by unscrewing and removing the screw cap, pouring salt into the salt space at the top of the appliance, and flushing through.

Water softening powder

An even simpler way to obtain small quantities of softened water for washing up, hair washing and baths, is to add a water softening powder such as Albright & Wilson's Calgon, to the water after it has been run into the bath, basin or washing machine. Calgon works by combining with the chemicals causing hardness and preventing them from having their unpleasant effects.

It should be remembered however that a hard water supply is not all loss, nor is soft water without its disadvantages.

If a plumbing system is of iron—rare nowadays but used very commonly between the wars—a hard water supply will

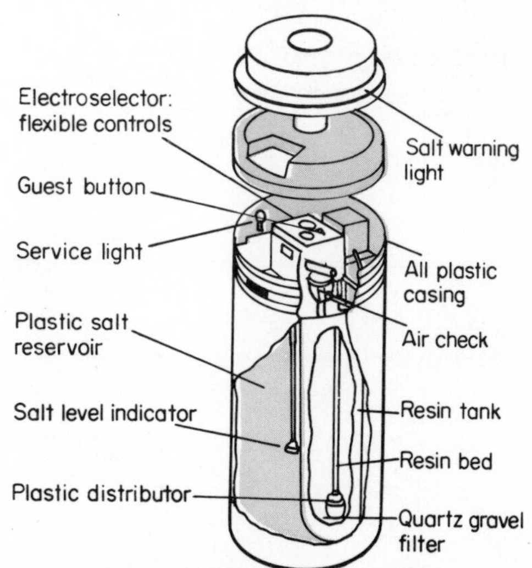

A modern automatic water softener such as the Sofnol Saturn has a large salt reservoir and automatically carries out the cycle of regeneration at regular intervals. Occasional replenishment of the salt reservoir is all that is required.

line the iron pipes with an eggshell layer of scale that will protect them against corrosion.

Danger to health

It has been suggested that, in certain circumstances, a soft drinking water supply can present a health hazard. As long ago as April 1969 the 'Lancet' (Britain's leading medical journal) published an article which claimed that deaths from cardio-vascular diseases were consistently higher in areas with a soft drinking water supply then in those with a hard supply.

Later studies have confirmed this claim. No one is quite sure of the reason. It could be that the chemicals causing hardness are needed as part of the defence mechanism of the human body. I think it more likely though that the explanation lies in the fact that soft water is a better solvent than hard, and more likely to pick up metallic contaminants, such as lead, from the pipes through which it passes.

Lead values greatly in excess of the accepted safety levels have been found in water from consumers' taps in soft water areas, particularly where the water had been standing in a lead pipe overnight or longer.

None of this need deter a householder from installing a water softener. Iron and lead pipes are not used in modern plumbing. Where they have been installed (in a hard water area where a softener is likely to be required) they will already have acquired an internal coating of eggshell scale.

To feel ***absolutely*** safe it might be wise to install the softener in the water supply pipe line ***after*** the branch taking the cold water supply to the kitchen sink. This will mean that you will still have hard water for drinking, cooking and the preparation of food. Soft water will be available for baths, washing and—from the hot tap over the kitchen sink—for washing up.

If you live in an area where the water supply is naturally soft, a sensible health precaution is to run off a few pints of water from the cold tap over the kitchen sink first thing in the morning before you fill the kettle for the pre-breakfast cup of tea.

Remember too, that hot water is a far more efficient solvent than cold. ***Never*** let late rising tempt you to fill the morning kettle from the hot water tap.

Chapter 10
Coping with frost

In the previous chapter we discussed the disconcerting property of water of taking into solution some part of practically any substance with which it comes into contact.

An even more significant characteristic of water is the fact that although, like everything else in nature, it expands when heated and contracts when cooled, it expands again when its temperature approaches 32°F (0°C) and it becomes ice.

From a world viewpoint this is extremely providential. If water ***contracted*** on freezing, then ice would sink to the bottom of rivers, lakes and oceans. Fish and other marine life could not survive. Since the sun's rays could never penetrate to the depths of the ocean to thaw the ice, any intelligent creatures that could survive in such a world would be familiar with water principally in its solid form.

From the rather narrower point of view of the householder, or the plumber, in earth's temperate zones, this property of water is a mixed blessing. Unless counter-measures are taken, the expansion of water into ice during a cold spell will stop the flow of water through pipes and cause them to burst.

The folk-myth that pipes 'burst with the thaw' dies hard. It is, of course, quite untrue. Pipes burst as the water within them expands and freezes. The burst becomes evident as the ice thaws and water starts to flow again.

The United Kingdom has had, at the time of writing, a succession of extremely mild winters. Vigilance will have relaxed. It is a safe bet that when the next prolonged cold spell comes (as it undoubtedly will) thousands of householders will find their plumbing systems frozen solid. With the thaw will come the remorseless drip of water through a thousand bedroom ceilings!

I hope that no readers of this book will be among those affected. Every householder should check his frost defences in the autumn; ***before*** the temperature drops to zero and icy north easters begin to whistle round the roof tops.

The external supply

Design safeguards against frost have already been referred to in Chapters 1 and 2. The external water supply pipe should be at least 0.82 m (2 ft 6 in) below the surface of the ground throughout its length and should be protected as it rises to the surface within the house. The rising main should rise to the cold water storage cistern against an internal wall. Boiler, hot water storage cylinder and cold water storage cistern should, as far as is possible, form a vertical column so that the vulnerable cistern may get the benefit of the ascending warm air.

Where pipe runs must, of necessity, be taken against an external wall they should be thoroughly lagged. An inorganic lagging such as fibreglass or foam plastic lagging units, is to be preferred. Make sure that the lagging extends behind the pipe. It is worse than useless to protect the pipe from the warm air within the house and leave it exposed to the cold wall!

The roof space

The roof space is a particularly dangerous area. If you have, in the interests of national fuel economy and your own rocketing fuel bills, insulated the spaces between the ceiling joists, you will have made your home warmer—and the roof space colder!

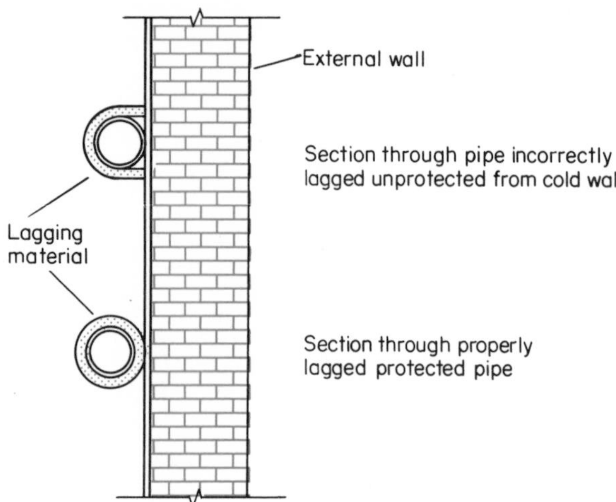

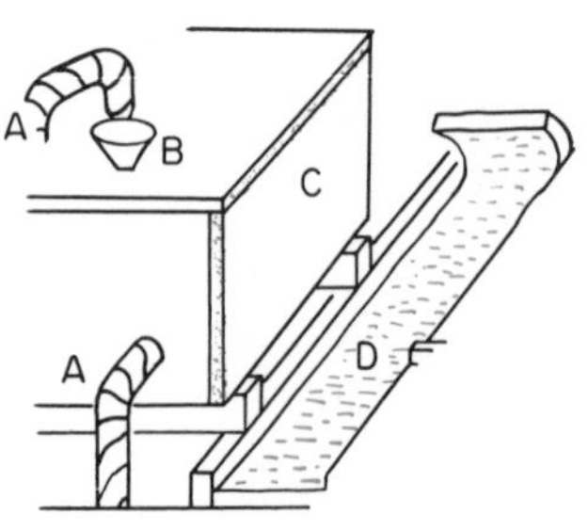

Always make sure that lagging extends behind the pipe that is to be protected. It is worse than useless to insulate the pipe from the warm air of the room and to leave it exposed to a cold external wall.

In the roof space all water pipes should be thoroughly lagged. A. The storage cistern should have a dust-proof cover into which is set a funnel to accept any water from the vent pipe B. The walls, but not the base, of the cistern should be lagged C, and the fibreglass mat laid between the rafters to conserve the house warmth should be omitted from immediately below the cistern D.

Pipe runs within the roof space should be as short as possible and should be thoroughly lagged. Don't forget to extend the lagging material to cover the tail of the ball-valve and all but the handles of any stop-valves.

The sides, but not the base, of the cold water storage cistern should be well lagged and the cistern provided with an insulating cover. Do not insulate the bedroom ceiling immediately below this cistern. There is something to be said for draping a 'curtain' of insulating material from the base of the cistern to the uninsulated ceiling below. This will funnel the slightly warmer air up to the base of the cistern.

The overflow or warning pipe from the cistern could permit icy draughts to penetrate the roof space. At one time it was the practice to provide a hinged copper flat at the external end of this pipe. This closed when the wind blew against it.

If you have this kind of protection apply a drop of oil to the hinge to make sure that it doesn't jam either open or closed.

Nowadays it is more usual to bend the internal end of this pipe inside the cistern so that it extends for an inch or so below the surface of the water. A water seal is thus provided that effectively prevents the entry of draughts. There are gadgets (such as the plastic Shire's Frostguard) that can be screwed on to the ends of existing overflow pipes to provide this protection.

Outdoor lavatories

Outdoor lavatories are very vulnerable in frosty weather. There are electro-thermal pipe lagging cables on the market which, plugged in to a power socket and switched on throughout frosty nights, will protect the supply pipe.

Provided that the lavatory is more or less draught proof, a 60 watt electric light bulb, switched on and suspended a few inches below the ball valve inlet (outside the cistern of course!) will supply sufficient warmth to protect against several degrees of frost.

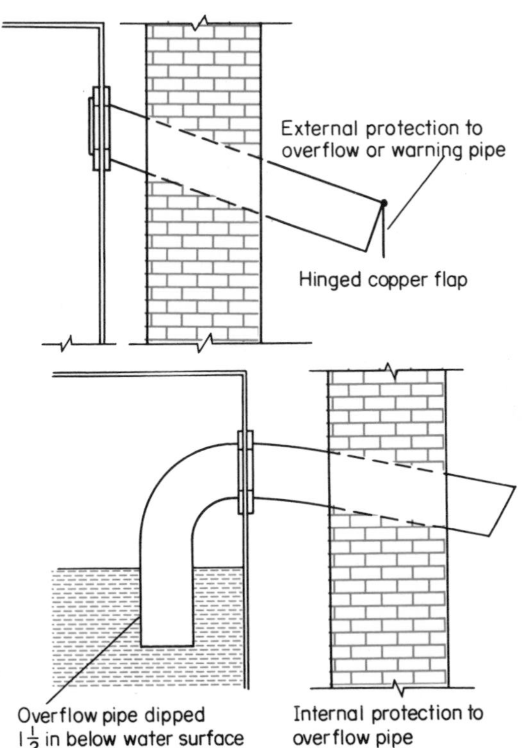

Icy draughts can be prevented from whistling up the overflow pipe into the roof space either by providing an external hinged copper flap or by dipping the internal end of the overflow pipe about 36 mm (1½ in) below the surface of the water to provide a trap. The latter method is to be preferred. Copper flaps are liable to jam open or closed.

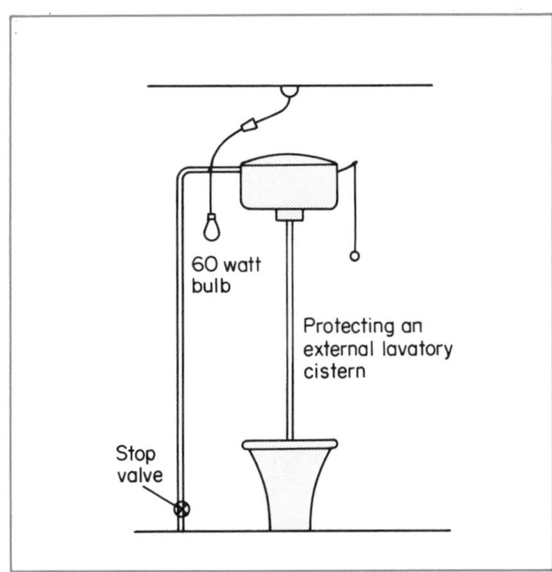

An external w.c. cistern can be protected from frost by suspending a 60-watt lamp bulb outside the cistern but below the inlet pipe. If this is switched on in icy weather it will, provided that the compartment is more or less draught-proof, give adequate protection.

Leaving the house unoccupied

It is important to appreciate that no amount of lagging (other than electro-thermal protection) will ***add*** heat to a plumbing system. All that it can hope to do is to conserve warmth already there. This point is particularly important as winter holidays become increasingly popular and many homes are left empty for two or three weeks at a time during the winter months.

For so long as a house is occupied, efficient lagging will protect against frost throughout a prolonged cold spell. Water flows into the plumbing system from the main at a temperature a few degrees above freezing point. Lagging reduces the rate of heat loss and the constant draw off and replacement of water in the pipes and storage cistern ensure that the domestic water supply remains above freezing point.

If the house is unoccupied during a prolonged spell of freezing weather even the most efficient lagging will only delay the eventual freeze-up. The fabric of the house chills off. Water stagnates in the supply pipes and becomes colder and colder. At last a plug of ice forms somewhere in a pipe-line—and quickly spreads throughout the system.

If you have a ***reliable***, automatic central heating system, turn it on at low level if you are going away for more than a few days at a time in the winter. Leave internal doors open so that warm air can circulate and remove the trap-door to the roof to allow some warm air to penetrate up there too.

Draining the system

The only other safe means of protecting your plumbing system, if you are absent from home for any length of time during which a cold spell might be expected, is to drain it completely.

Turn off the main stop-cock and drain from the drain-cock immediately above it. Open all hot and cold taps and leave open. Connect one end of a length of hose to the drain-cock beside the boiler (or at the base of the cylinder if you haven't a boiler) and take the other end to an outside gully. Open up the drain-cock and leave until no more water flows.

Human memory is fallible. Having done this write PLUMBING SYSTEM DRAINED —DO NOT LIGHT BOILER UNTIL REFILLED on a card and prop it on the boiler.

Incidentally, when refilling, you can reduce the risk of air-locks forming by connecting one end of your hose to the cold tap over the kitchen sink and the other end to the boiler drain-cock. Open both up and the system will fill *upwards*, driving air in front of it.

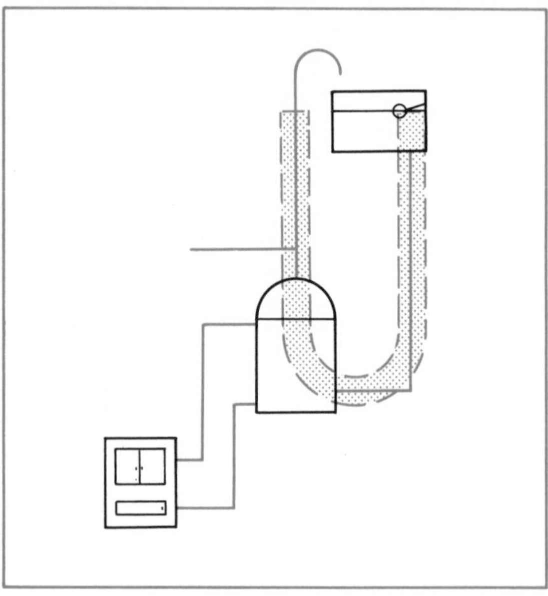

The open ended U-tube formed by a cylinder storage hot water system is the first, and strongest, defence against the possibility of a boiler explosion.

Boiler explosions

Many people, particularly the elderly, worry unnecessarily about the risk of a boiler explosion during icy weather. This may lead them to let their boiler fires out at night which is the worst possible course of action.

Boiler explosions are, in fact, extremely rare but, when they do occur, their effects are so catastrophic that it is not surprising that they should be a source of anxiety.

Domestic hot water systems form a kind of extended U-tube with the vent pipe and the cold water storage cistern as the two open ends. Provided this U-tube remains unobstructed there is no risk of an explosion. The safety valve usually situated by the boiler provides a final line of defence against the possibility of an obstruction occurring.

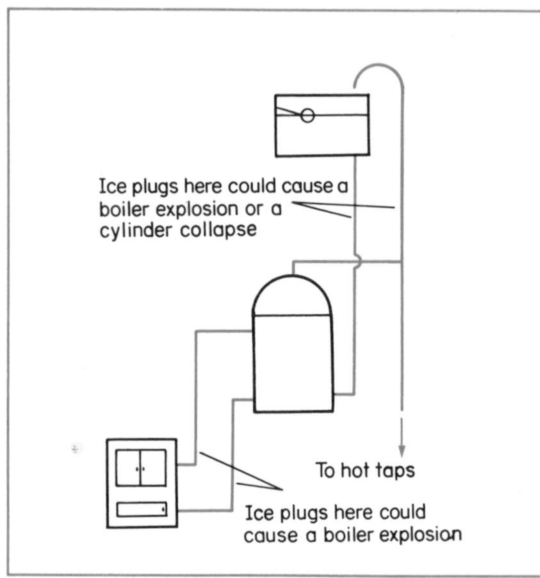

A boiler explosion can occur if the boiler fire is lit after ice plugs have blocked the U-tube at the points indicated.

Typically, boiler explosions occur when a family returns from a winter holiday without having taken the precautions suggested in this chapter. During their absence, plugs of ice have formed in the vent pipe and in the supply pipe from the cold water storage cistern to the hot water cylinder. There could even be ice plugs in the flow and return pipes between cylinder and boiler.

The boiler fire is lit and the water in the boiler heats up. It cannot circulate and quickly overheats, rapidly reaching a temperature well above boiling point. It cannot however boil and turn to steam because it is confined to the limited space of the boiler and perhaps a few feet of circulating pipe.

Pressure within the boiler builds up and, eventually, something gives—releasing the internal pressure. Instantly, the superheated water turns to steam—with many thousand times the volume of the water from which it was produced—and the boiler explodes like a bomb, usually with equally lethal results.

Cylinder ***implosion***, or collapse, during icy weather is a rather more common phenomenon.

This can happen when a boiler fire, kept alight all day, is let out at night. Ice plugs form in the upper part of the vent pipe and the cold water supply to the cylinder, blocking the U-tube. In the meantime the water in the system begins to cool and also contract.

Copper hot water storage cylinders are manufactured to withstand considerable internal, but very little external, pressure. Collapse, like a paper bag, often occurs first thing in the morning when the householder attempts to draw off some hot water—the final straw that breaks the camel's back!

Sometimes, when the ice-plugs thaw and water flows back into the cylinder again, the resumed internal pressure will restore the cylinder undamaged to its former shape; but I wouldn't guarantee this.

How to tackle a freeze-up

Supposing, despite all your precautions, you still get a freeze-up. You will know about it because water will cease to flow from one or more of the taps.

Take immediate action. Find the ice-plug and thaw it out before it has a chance to spread through the system. If water is still running from the cold tap over the sink but is not reaching the cold water storage cistern, then the ice-plug must be in the rising main between the sink supply branch and the ball-valve feeding the cistern.

Strip off the lagging and apply heat to the pipe; cloths soaked in hot water and then wrung out are a good way to do this. Don't use a blow torch among the dry timbers of the roof space. An electric hair-dryer, or even a vacuum cleaner operating in reverse, provide means of directing a stream of warm air to inaccessible lengths of pipe.

If the freeze-up is tackled quickly it will be cleared easily. The copper of which modern water supply pipes are made is a good conductor of heat. Heat applied to the pipe will be conducted along it to clear an ice plug that might be several feet away.

Repairing a burst pipe

And supposing you get a burst pipe? The first indication is likely to be water dripping from a ceiling.

Once again, immediate action is called for. Turn off the main stop-cock and open up all the taps to limit the amount of damage. Only then should you look for the site of the burst.

If you have copper tubing joined by non-manipulative compression fittings or by soldered capillary joints, the chances are that expansion of the ice will have simply forced the joints open. The joint can be remade (see Chapter 11).

A lead pipe will, quite probably, have split under pressure from the ice.

The orthodox method of repairing a burst lead pipe is to cut out a length of pipe extending 9 in to 1 ft on each side of the burst, and to insert a new length, joining to the old pipe with wiped soldered joints.

This is, in my opinion, strictly a job for a professional plumber.

However the householder can make a 'temporary' repair (which may last as long as a permanent one!) using one of the epoxy resin repair kits (Isopon, Plastic Padding, Holts Fibreglass repair kit and so on) that are available from all d.i.y. shops.

Make sure that the pipe is clean and dry and knock the edges of any split together. Rub down with coarse abrasive paper to form a key. Mix up the epoxy resin filler and hardener according to the makers' instructions and 'butter' over the area of the leak and for a few inches on either side. While the filler is still plastic, bind a fibreglass bandage round the buttered area and, finally, apply another coating of resin filler.

A repair of this kind, which will permit you to have the pipe in use again within a few hours, may not meet with the approval of your local water authority. I think though that even they would agree that it is preferable to letting the water leak and waste away until you could obtain the services of a professional plumber.

Chapter 11
Plumbing techniques for the householder

This chapter, which may well be of most interest to the d.i.y enthusiast, has been deliberately placed at the end of this book. Before attempting any plumbing operation it is essential that the householder should have a thorough grasp of the principles involved.

No-one who has read the preceding chapters would make such elementary—and serious—errors as, for instance, taking the cold supply to a conventional shower direct from the main, connecting a rim supply bidet to existing bathroom hot and cold water supplies or disconnecting the flow and return pipes from a hot water storage cylinder without having drained the system ***from the drain-cock beside the boiler.***

In this chapter I shall deal with plumbing techniques—joining and bending lengths of pipe and connecting them to taps and other fittings—which, in my opinion, are well within the scope of the determined home owner. Making wiped soldered joints between lengths of lead pipe, bronze welding, lead and zinc roof work are important aspects of the professional plumber's skill but are not tasks for the inexperienced.

The development of copper, stainless steel and plastic tubing has provided the amateur with materials that he can handle safely and effectively.

Copper tubing

Copper tubing is undoubtedly the most common plumbing material used in post-war homes. Its development, and almost universal adoption, has probably been the biggest single factor in bringing home plumbing within the scope of the home handyman.

The sizes of copper tubing most likely to be used in domestic plumbing are 15 mm, 22 mm and 28 mm. The Imperial equivalents of these sizes are ½ in, ¾ in and 1 in respectively.

A check with a rule will reveal that these are not literal translations from metric to Imperial measurements. The reason for the apparent discrepancy is that Imperial sizes are the measurements of the internal diameter of the tube. This means that all tubes of equal capacity, no matter of what material they are made, are of the same Imperial size. The metric measurement is of the ***external*** diameter; Don't ask me why!

Joining copper tubing

Copper tubing may be joined by means of non-manipulative (Type A) manipulative (Type B) compression joints and fittings or by means of soldered capillary joints.

Non-manipulative compression joints are the easiest means of joining copper tubing for the amateur with no previous experience and a minimal tool kit. The only essential tools are a hacksaw, a rasp, a spanner of the appropriate size and a wrench. If, however, you have a fairly large plumbing project in hand, a tube cutter incorporating a reamer is a good investment.

Non-manipulative compression joints have three essential features; the joint body, a soft copper ring or 'olive' and a cap nut. To make the joint, the tube end must be cut absolutely square and all internal and external burr removed. This is where the tube cutter with reamer comes in useful.

Loosen the cap nut of the compression fitting (with most makes there is no need to remove it) and thrust the tube end in as far as the tube stop. Hold the body of the joint firmly with the wrench and tighten the cap nut with the spanner. This action compresses the soft metal ring against the outside surface of the tube to provide a secure and watertight joint.

Provided that you use a spanner, and not a wrench, to tighten the cap nut, it is practically impossible to overtighten. Many plumbers add a smear of boss white or similar jointing material to the tube end and to the inside of the compression fitting. It should not be necessary to do this but it does ensure a watertight joint first go.

Any builders' merchant will let you browse through his illustrated catalogues of compression joints and fittings. You will find there fittings for every conceivable purpose—straight couplings, reducing couplings, bends, tees and so on.

Connecting new metric tubing to existing imperial tubing presents no serious problem with this kind of joint. 15 mm and 28 mm compression fittings are exactly interchangeable with ½ in and 1 in fittings respectively. Thus, you can, for instance, fit a 15 mm compression tee joint into a run of ½ in tubing to take a new 15 mm branch supply to a wash basin or lavatory cistern.

22 mm fittings are not ***exactly*** interchangeable with ¾ in fittings. The body of a 22 mm joint can be used with ¾ in tubing but a special copper ring and cap nut are needed to ensure a watertight joint. Your supplier should be able to fix you up with these.

The home plumber isn't really very likely to need to use manipulative (Type B) compression fittings. Many Water Authorities however require this kind of fitting for use underground.

With a manipulative compression fitting the cap nut must first be unscrewed from

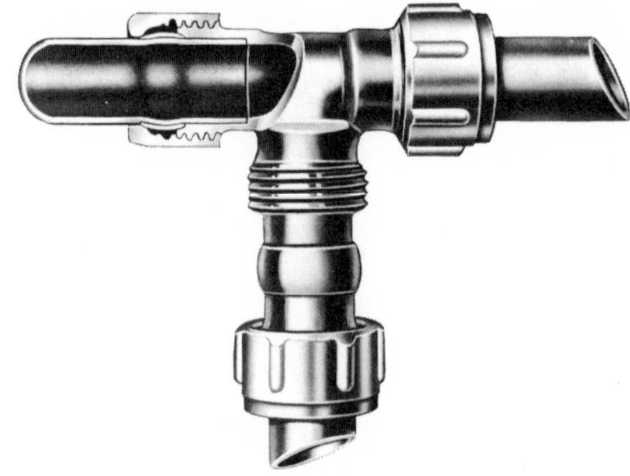

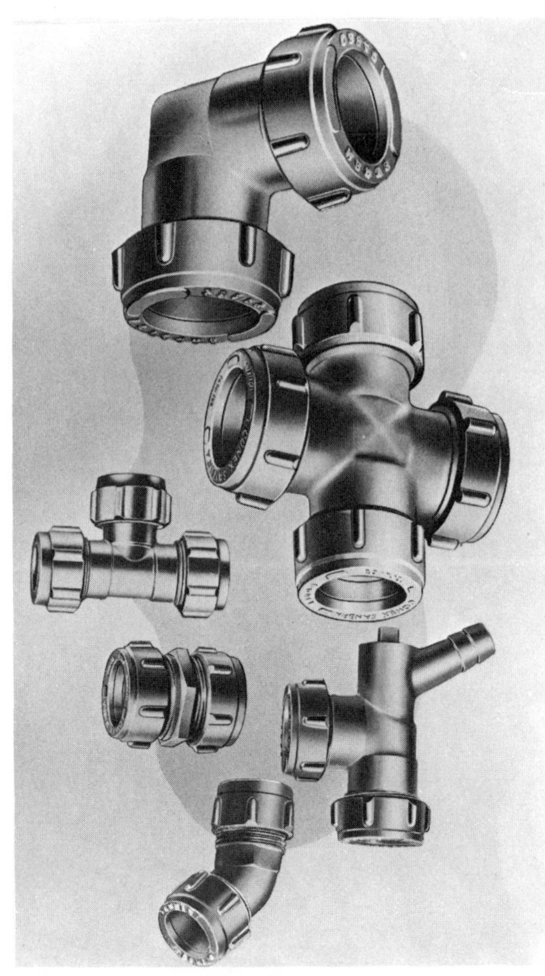

When the cap-nut of a compression joint is tightened the soft copper ring or olive in compressed against the outer wall of the copper or stainless steel tube to provide a watertight joint. Compression fittings, like those illustrated, can be used for a variety of plumbing jobs (Courtesy of Conex-Sambra Ltd)

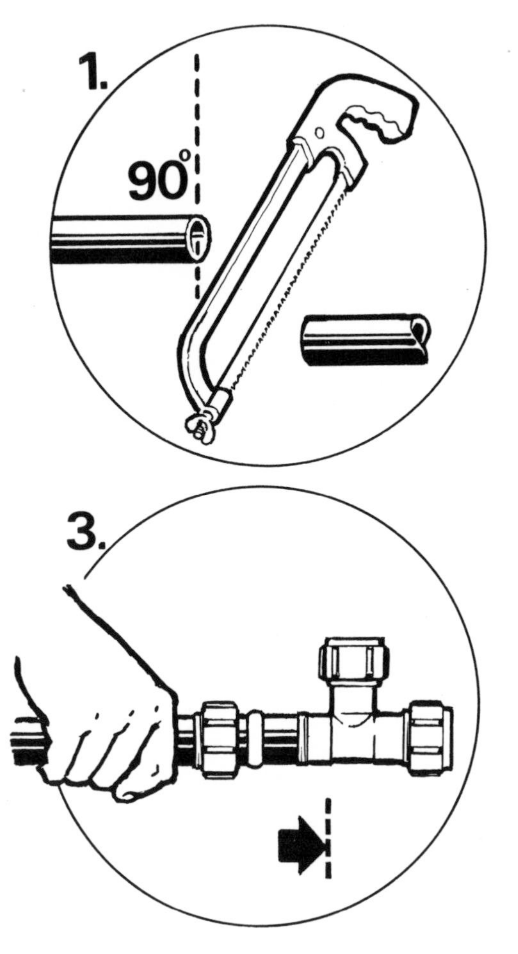

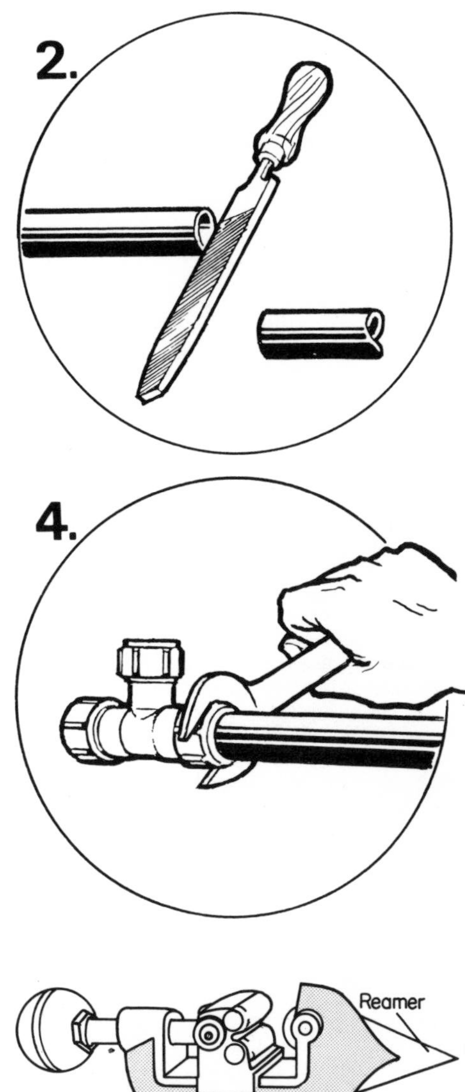

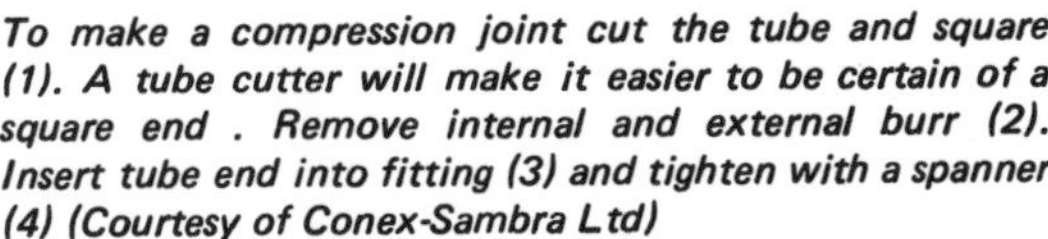
To make a compression joint cut the tube and square (1). A tube cutter will make it easier to be certain of a square end . Remove internal and external burr (2). Insert tube end into fitting (3) and tighten with a spanner (4) (Courtesy of Conex-Sambra Ltd)

A tube cutter with reamer is shown on the right

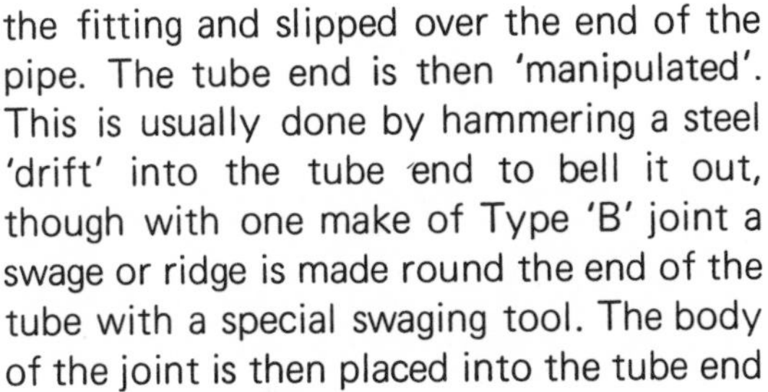
the fitting and slipped over the end of the pipe. The tube end is then 'manipulated'. This is usually done by hammering a steel 'drift' into the tube end to bell it out, though with one make of Type 'B' joint a swage or ridge is made round the end of the tube with a special swaging tool. The body of the joint is then placed into the tube end and the cap-nut tightened up hard. With this kind of joint the use of boss white or some other waterproofing compound is recommended.

It can be seen that a Type 'B' fitting cannot be dismantled in the same way that a Type 'A' fitting can—and it cannot be pulled open by expansion due to frost.

Straight connector
(copper to copper)

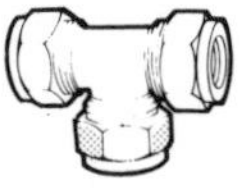
Tee junction

Slow bend

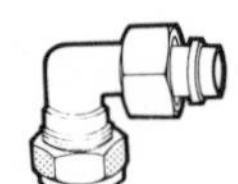

Bent tap or ball-valve
connector

Wall plate elbow for
outside tap

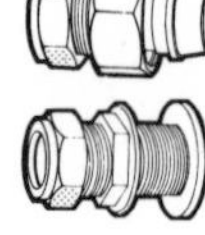

Connection to
storage cistern

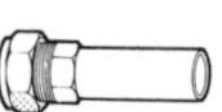
Copper to lead joint

Compression fittings are made for all purposes. Here is a small selection from the Prestex range:

A straight connector for joining two lengths of copper or stainless steel tubing.

A tee junction for taking a branch water supply from a main supply pipe.

A slow bend.

A tap connector with compression joint at one end and cap and lining for connection to the tail of a tap at the other.

A bent tap or ball-valve connector. This is very useful for connecting the ball valve to a cold water storage cistern. The rising main is taken up the outside of the cistern vertically and the bent connector fitted onto its end with the cap and lining joint in position to connect to the threaded tail of the ball valve.

A connection to a storage cistern.

A wall plate elbow designed to take a bib-tap for a garden water supply. PTFE thread sealing tape should be bound round the tail of the tap before screwing it home into the female thread of the wall plate elbow.

A copper or lead joint. The 'lead end' of this joint must be connected to the lead pipe by menas of a soldered joint.

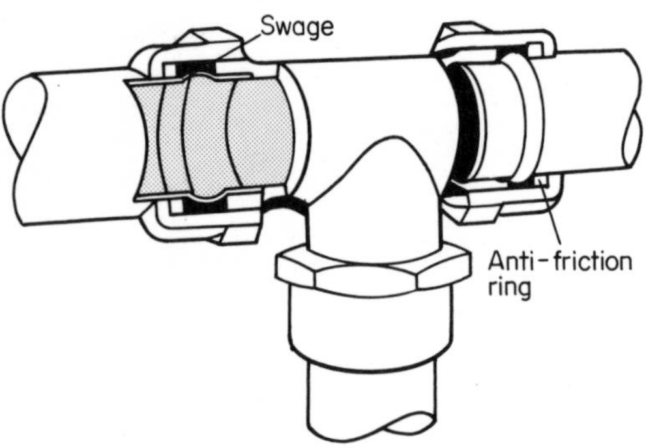

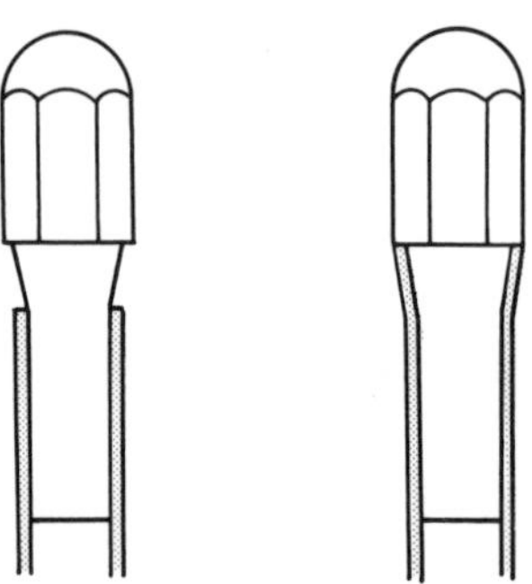
A steel drift opens up the tube end

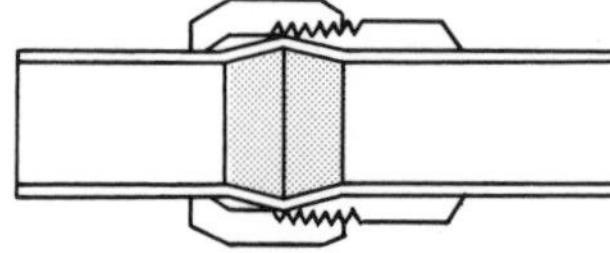

Manipulative, Type B, compression joints are used for underground water supply pipes. The cap nut is slipped over the tube end which is then 'manipulated' either by opening the end with a steel drift or, as in the Kingley coupling, by forming a swage with a swaging tool. Manipulating joints cannot pull apart and cannot, of course, be readily dismantled for re-use.

Making a soldered joint

Soldered capillary joints and fittings are smaller, neater (and cheaper) than compression fittings. Using them is well within the scope of the determined handyman. Their effectiveness depends upon capillary action—the force which makes liquids (in this case liquid solder) flow into any confined space between two smooth, solid surfaces.

As with compression joints there are two kinds of capillary fitting; the integral ring and the end-feed fitting. Integral ring fittings, often called 'Yorkshire fittings' (though this is the name of just one well-known brand), have a ring of solder, sufficient to make the joint, incorporated in the fitting itself. With end-feed fittings solder to make the joint has to be added with solder wire.

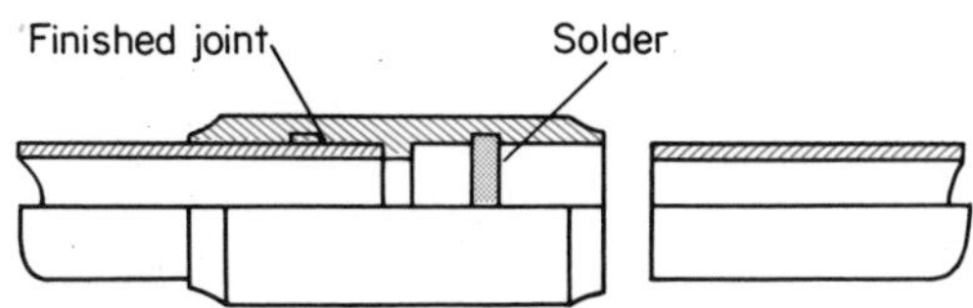

Integral ring soldered capillary joints incorporate airing of solder sufficient to make the joint.

End-feed fittings are identical except that they do not incorporate a solder ring. After heating, solder is fed into the mouth of the joint from solder wire.

Absolute cleanliness is the key to success in making any kind of capillary soldered joint.

As with a compression joint the tube ends must be cut square and all trace of internal and external burr removed. Clean the tube end and the internal bore of the fitting thoroughly with steel wool and smear an approved flux onto the tube end and internal bore.

Insert the tube into the fitting to the tube stop. All that now has to be done with an integral ring fitting is to apply the flame of a blow-torch—a butane torch is perfectly satisfactory for this purpose—first to the tube and then to the fitting. The joint is made when the solder melts and appears as a ring all round the mouth of the fitting.

End-feed fittings are, naturally, cheaper and are not a great deal more difficult to use. The tube end and the fitting are prepared and fluxed as with an integral ring joint. Bend over a length of solder wire—about ½ in for a 15 mm fitting, ¾ in for a 22 mm fitting, and so on—and, after preliminarily heating the tube and fitting, feed the bent-over length of solder into the end of the joint. Once again, the joint is complete when all the bent-over length of solder has been melted and drawn into the joint and a bright ring of solder is apparent all round the mouth of the fitting.

Making an integral ring 'Yorkshire' soldered capillary joint.

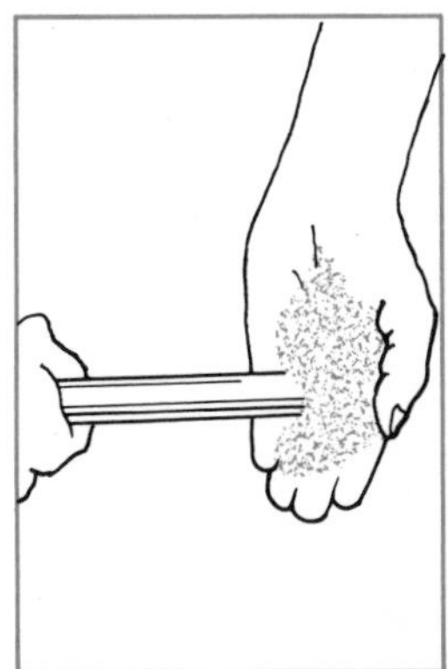

1. clean end of tube and bore of fitting with steel wool

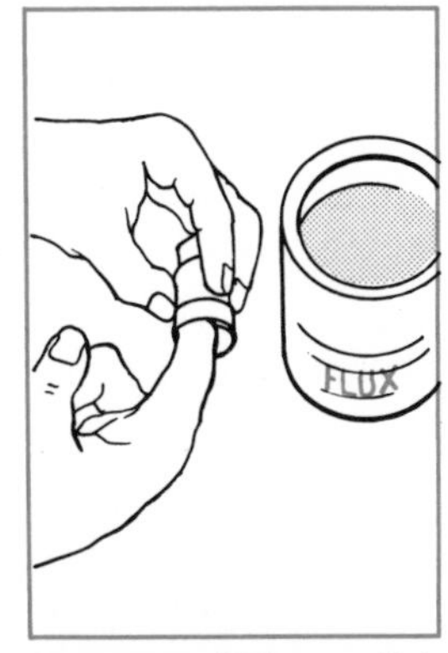

2. flux bore of fitting and tube end. With phosphoric acid flux use a brush!

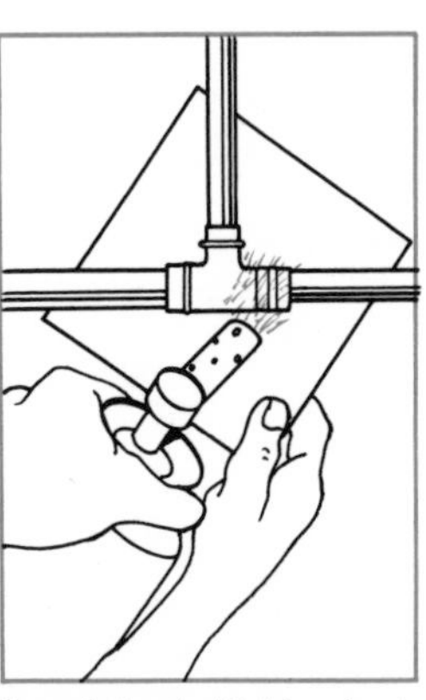

3. apply heat with blow torch (note asbestos sheet behind fitting)

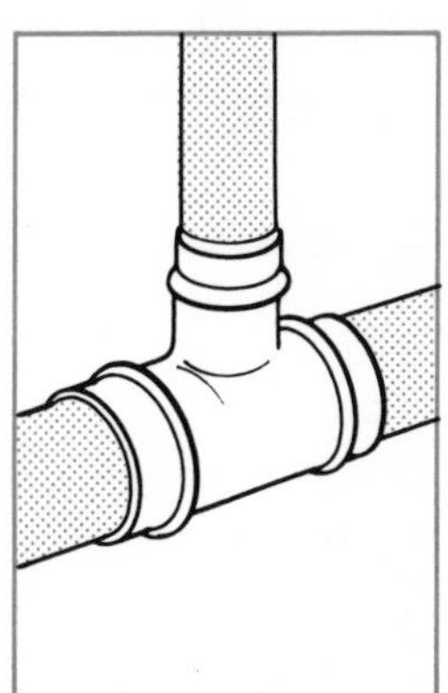

4. leave completed joint to cool

Once the joint is made do not disturb until the solder has set and the fitting is cool enough to touch. Where—and this is usually the case—more than one joint is to be made with one capillary fitting, (for instance, the two ends of a straight coupling or the three ends of a tee junction) it is best to make all the joints at the same time. If this cannot be arranged, a piece of damp cloth should be wrapped round any joint already made to prevent the solder from melting.

Always bear in mind the fire risk involved when using a blow torch to make capillary joints. It is all too easy to become so engrossed in making a first class joint that the smouldering timber behind the pipework goes unnoticed until too late! Interpose a sheet of asbestos between the pipe on which you are working and the skirting behind it. Be particularly careful when working amid the bone-dry timbers of the roof space.

Unlike compression joints, 15 mm, 22 mm and 28 mm capillary joints ***cannot*** be used with ½ in, ¾ in and 1 in tubing. Capillary action requires a more critical fit than does a compression joint.

Imperial to metric couplings and tee joints are manufactured but a simple way out is to use a compression fitting for the actual connection between old Imperial and new metric tubing and then to carry on using metric capillary joints.

Making a watertight joint

Every manufacturer supplies compression and capillary fittings with threaded ends (either male or female) for connection to galvanised steel tubing, cylinder tappings or to take the back-nuts securing pipework to storage cisterns. Screwed joints may be made watertight by binding PTFE plastic thread sealing tape round the male thread.

PTFE is sold in rolls, rather like the familiar rolls of surgical tape. Tear off an appropriate length, bind round the thread and screw home.

Although there are no technical difficulties involved in connecting new lengths of copper tubing to an existing galvanised steel plumbing system, the risk of electrolytic corrosion (see Chapter 1) should be borne in mind. It is always unwise to use copper and galvanised steel in the same plumbing system; particularly for hot water services. Where additions are to be made to a galvanised steel plumbing system it is safer to use the stainless steel tubing referred to later in this chapter.

Connecting taps and ball-valves

Taps and ball-valves are normally connected to copper (or any other) tubing by means of swivel tap connectors or 'cap and lining' joints.

These useful little fittings are inserted into the tail of the tap or valve and the nut is tightened up on to the thread of the tail. They are normally supplied with a fibre washer that makes the use of PTFE tape unnecessary.

Connecting new copper tube to lead pipes

One of the most difficult tasks likely to confront the amateur plumber is the connection of new copper tubing to an existing lead pipe. Typically a householder wants to replace a leaky and out-of-date lead plumbing system with a copper one—and he has to begin by making a connection to the lead rising main protruding from the kitchen floor.

All manufacturers of compression and capillary fittings include lead-to-copper joints among their range. Unfortunately the lead end of this fitting has to be connected to the lead pipe with some kind of soldered joint.

The conventional, approved and professional method of doing this is with a wiped soldered joint. Few amateurs are likely to make a success of such a joint at their first, or even their second, attempt and the vertical position in which the joint is likely to have to be made makes the task even more difficult.

I would strongly advise the amateur to seek professional help for this part of the job at least. Once the lead-to-copper connection has been made he will have no difficulty in continuing with his plumbing project.

There are two lead-to-copper joints which are within the capacity of the householder but there are snags about both of them. The first, the Staern or soldered spigot joint, provides a kind of *in situ* capillary joint. It is little used in the trade nowadays and the home plumber may find difficulty in getting hold of the necessary tools.

The Staern joint provides — if you can obtain the tools required — a straightforward means of connecting lead tubing to the brass end of a lead-to-copper compression joint.

a A cutting tool is used to chamfer the mouth of the lead pipe.

b The end of the lead pipe is opened out with a mandrel.

c The end of the lead pipe is then shaped inwards with a coning tool.

d The brass tail is then tinned and inserting after being fluxed. The result might be described as an 'in situ' soldered capillary joint.

The method of making the joint is shown in the diagram. A special cutting tool is inserted into the end of the lead pipe and rotated to produce a chamfered lip. The mouth of the pipe is then opened with a mandrel and a hollow hardwood cone is tapped down over the opened out pipe end to shape the sides inwards.

Next clean and rasp the end of the brass or gunmetal lead-to-copper joint and 'tin' the surface with solder. Smear the tinned end of the fitting with flux and insert into the end of the lead pipe. Apply a blow-torch flame to the brass fitting and bring down to heat ***gently*** the end of the lead pipe. Feed solder into the lip of the lead pipe from whence it will flow, by capillarity, into the space between the spigot of the brass fitting and the lead socket.

As with a conventional capillary joint a ring of bright solder will appear round the mouth of the joint. Continue heating gently until bubbles of flux cease to rise. Wipe off surplus solder while still plastic to give a neat finish.

The other joint, the cup and cone joint, is easily made but is unlikely to meet with the approval of the local Water Authority, certainly not for a pipe carrying water under mains pressure.

To make this joint the end of the lead pipe is opened out with a hardwood cone until the spigot of the brass fitting can be

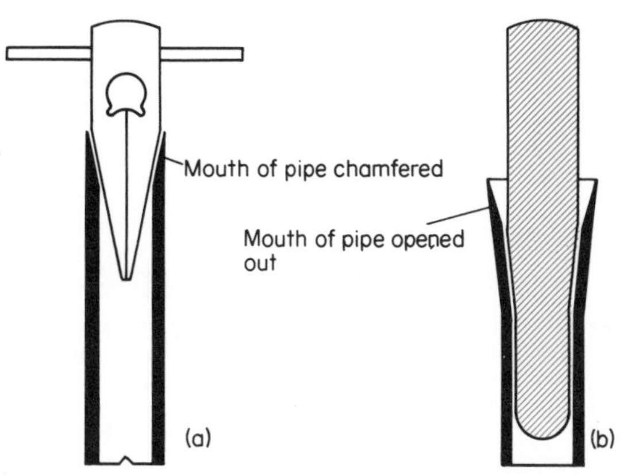

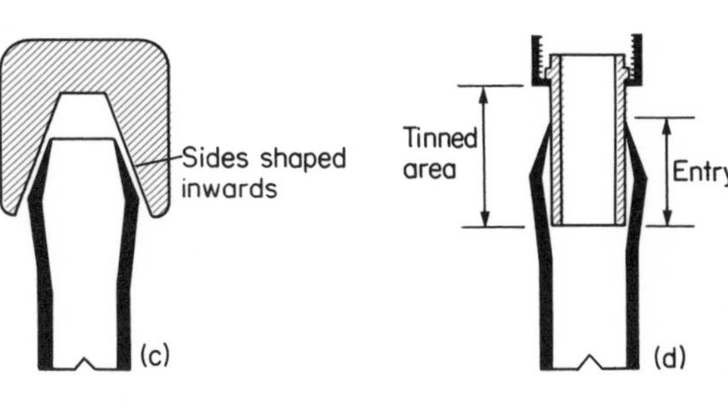

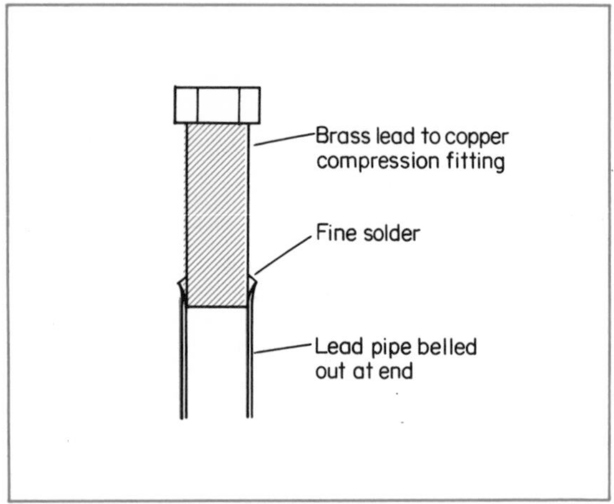

The cup and cone joint illustrated is an easy method of joining brass to lead. There is no reason why a joint of this kind should not be used for a waste pipe, which does not carry water under pressure, but it will not meet Water Authority approval for a water supply pipe.

accommodated to a depth equal to half its diameter—¼ in for a 15 mm fitting. Rasp, tin and flux the spigot end of the fitting. Insert into the belled out end of the lead pipe and run fine solder into the space between the belled end and the brass spigot.

As well as being able to join lengths of pipework the amateur plumber needs to know how to negotiate bends. All makers of compression and capillary include a variety of bends in their ranges.

However, copper tubing can, with the aid of bending springs, be bent by hand to easy bends and, for a major plumbing job, the use of this technique can save a good deal of money.

Pipes of up to 28 mm (1 in) *can* be bent cold by hand but the amateur would be wise to limit himself to easy bends in 15 mm (½ in) and 22 mm (¾ in) tubing. The purpose of the bending spring is to support the walls of the pipe which would otherwise collapse as the bend is made.

This is the method: Use a spring of the correct size, grease to facilitate easy withdrawal, and insert into the tube to the point at which the bend is to be made. Bend over the knee. Best results will be obtained by slightly overbending at first and then bringing back to the required curve.

To withdraw the spring insert a tommy bar through the metal loop at the end, turn clockwise to reduce the diameter of the spring, and pull. ***Never*** be tempted to 'dress', hammer or otherwise interfere with the tube until the spring has been withdrawn. If you do you will find it to be irretrievably jammed within the tube.

Stainless steel tubing

It has been a source of surprise to me that stainless steel tubing, which has been available in this country for some ten years, has not proved to be more popular than it has. It offers many advantages both to the professional and to the amateur plumber.

As a home produced product it has a relatively stable price which has been consistently below that of copper tubing.

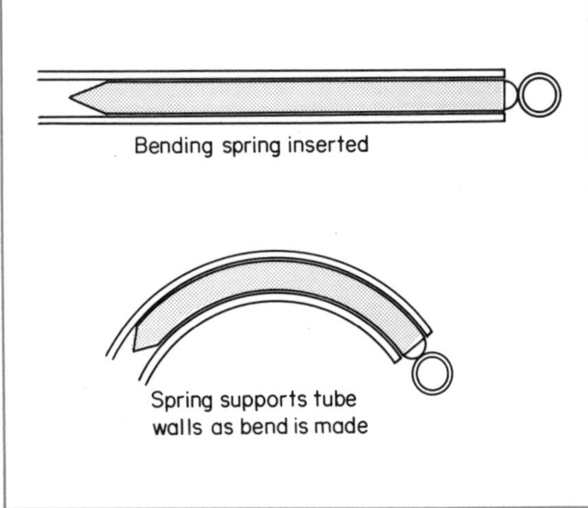

Bending copper tubing with the aid of a bending spring. The spring is greased and inserted (above) to the point at which the bend is to be made. The tube is then bent over the knee (below), first overbending and then bringing back to the curve required. To remove the spring insert a tommy bar into the loop at the end. Twist clockwise to reduce the diameter of the spring – and pull.

It has an attractive appearance and needs no decoration. Above all, it can be used in conjunction with galvanised steel tubing (provided that this is not already rusting) or copper tubing, without risk of electrolytic corrosion.

It is no more difficult to use than copper tubing and the same methods of jointing and manipulation can be adopted.

Joining stainless steel tubes

Stainless steel tubing can be joined by means of either Type A or Type B compression fittings or by means of either integral ring or end-feed capillary fittings. There are however one or two points of technique that must be borne in mind.

Preparation of the tube ends for any kind of jointing is the same as with copper tubing. Stainless steel tube can be cut with a tube cutter or with a hacksaw but, with stainless steel, a hacksaw—ideally a high speed steel hacksaw blade with 32 teeth per inch—is to be preferred. This is particularly important if Type B (manipulative) compression fittings are to be used. The use of a tube cutter work-hardens the tube ends and there would be a tendency for the end to split when opened out with a drift.

Apart from this, both types of compression joint are used with stainless steel exactly as with copper. Since stainless steel is a harder material a little more pressure needs to be applied when tightening the cap nut in order to ensure a watertight joint.

When joining stainless steel tubing with either integral ring or end-feed capillary fittings it is important that a flux based on ***phosphoric acid*** should be used. A suitable flux is Phosphor-Rite, marketed by Southern Cross Fluxes, but others are available. ***On no account should you use a chloride based flux.***

In making a capillary joint with stainless steel a gentle flame from your blow-torch should be applied to the fitting itself—***not*** to the tube.

Many manufacturers provide a, possibly limited, range of chromium plated compression and capillary joints and fittings intended for use with stainless steel tubing.

Stainless steel tubing is harder than copper tubing and is consequently more difficult to bend. The amateur might be well advised to use compression or capillary bends for all changes of direction but it is possible to make easy bends in 15 mm (½ in) stainless steel tubing using the bending spring technique already described.

Other pipe materials that may be used, or encountered, by the home plumber are plastics (polythene, polyvinyl chloride—PVC or vinyl for short—and polypropylene) and pitch fibre.

Polythene tubing

Black polythene tubing enjoyed quite a vogue with amateur plumbers a few years ago because of the long lengths in which it was obtainable, the ease with which it could be connected to copper tubing and its built-in resistance to frost.

It has disadvantages however. It cannot be used for hot water supply; though it is satisfactory for waste pipes which do not carry hot water under pressure. It has a rather thick, clumsy appearance and its tendency to sag necessitates continuous support on horizontal runs.

A very important use of polythene is as a material for underground supply pipes; perhaps taking a supply to a garage or to a stand-tap at the bottom of a large garden. Neither its appearance, nor its need for support, present problems when used in this way. The long lengths in which it is obtainable, eliminating underground joints, and its in-built insulation give it very real advantages.

Polythene tubing is joined with non-manipulative compression fittings similar to those used with copper tubing. Polythene has not, at the time of writing, been metricated and is still sold as ½ in, ¾ in or 1 in internal diameter. When ordering compression fittings for polythene tubing it is wise to take a short length along to the builders merchant to make sure that you get the right size fittings.

Because polythene is a relatively soft material a metal insert, provided by the manufacturer of the compression fittings, must be inserted into the tube end as the joint is made.

The correct procedure is to unscrew the cap nut of the compression joint and slip it, followed by the copper ring or olive, over the end of the tube. Push the metal insert into the end of the tube. Insert the tube end into the body of the fitting as far as the stop and tighten up the cap nut. Tighten as far as possible with the fingers and then give one and a half to two turns with a spanner.

Easy bends can be made cold with polythene tubing provided that the bend is then firmly secured. For making permanent bends the tubing must first be heated. A professional plumber would probably do this by ***very*** gently playing a blow-lamp flame over the length to be bent. An amateur might be better advised to immerse the length for ten minutes in water that is kept boiling for that period.

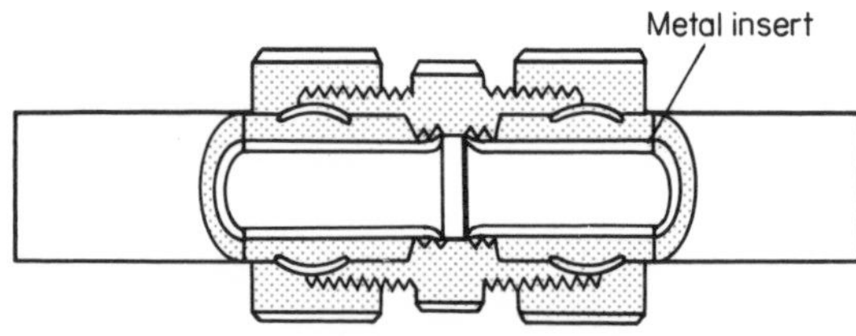

Polythene tubing may be joined by means of Type A non-manipulative compression joints but a metal insert is required in the end of the polythene tubing to protect it from collapse when the cap-nut is tightened. A polystantor compression joint is shown.

Polyvinyl chloride (PVC or Vinyl) tubing

PVC is nowadays very widely used for above and underground drainage and for roof drainage. It may also be used for cold water supply and provides a cheap and quickly assembled means of providing all domestic cold water services.

PVC cannot be used for hot water under pressure. Because of this there are two, nominally cold water supply pipes which should never be of PVC. These are the cold water supply pipe from the cold water storage cistern to the hot water storage cylinder and the cold water supply pipe from the feed and expansion tank of an indirect hot water system. Metal pipes should always be used in these positions as the water in these pipes can become very hot at times.

PVC tubing can be joined either by solvent welding or by ring seal jointing.

Solvent welding must be used for cold water supply pipes. For waste and drainage pipes a mixture of the two methods is often used—solvent welding for small diameter waste branches and ring seal jointing for the larger diameter stack and drain pipes.

To make a solvent welded joint the tube end must be cut off squarely with a hacksaw and all swarf and burr removed from internal and external surfaces.

With a file or fine toothed rasp chamfer the outer edge of the pipe end to an angle of 15° or 20°. Insert into the socket and mark with a pencil the extent to which the pipe end enters the socket. Roughen the pipe end and the interior surface of the socket with fine abrasive paper or cloth and degrease these surfaces using a cleaning fluid approved by the manufacturers and clean absorbent paper.

Using a brush, apply an even coat of solvent cement to both pipe end and fitting in lengthwise strokes. The pipe end

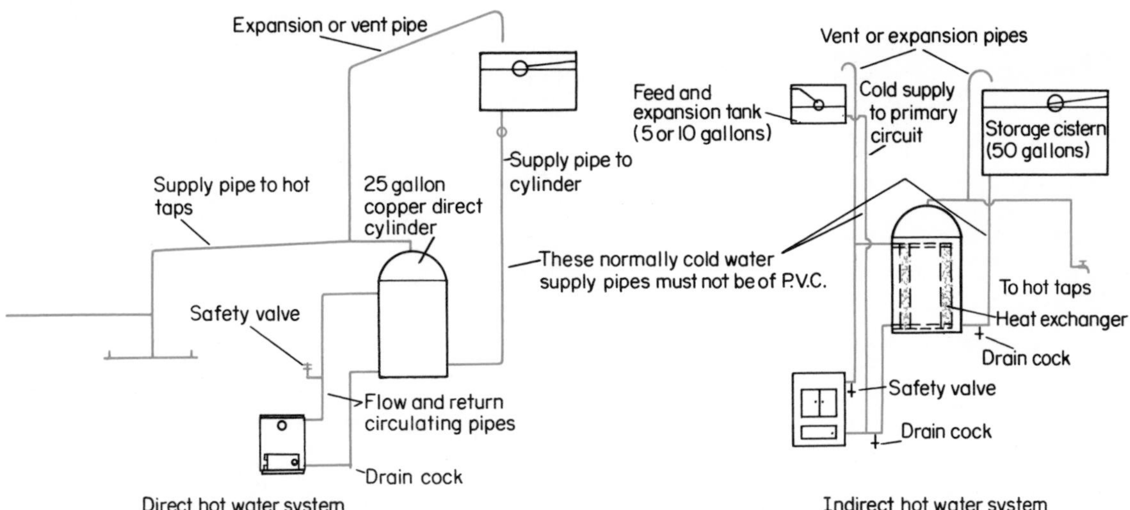

PVC tubing may be used for all cold water supply pipes but should not be used for nominally cold pipes which may, in fact, sometimes get very hot. PVC should not be used for the cold water supply to the hot water cylinder of a direct hot water system (top left). It should not be used for the cold water supply to the cylinder or the cold water supply to the primary circuit in an indirect system (top right).

should have a thicker coating than the socket surfaces.

Immediately push the socket on to the pipe without turning, hold in position for about fifteen seconds and then remove surplus cement. The joint should not be disturbed for approximately five minutes and should not be put into use for twenty-four hours.

Preparation for ring seal jointing is similar. Use a fine tooth woodsaw or a hacksaw to cut the—usually—larger diameter pipe. A useful tip, to ensure an absolutely square cut, is to place a sheet of newspaper over the pipe and to bring the edges together beneath it.

Draw a line round the cut end of the pipe 10 mm from the end and chamfer back to this line with a rasp or other shaping tool. Insert the pipe end into the socket and mark the insertion depth, making an allowance for expansion of 10 mm between the end of the pipe and the bottom of the socket. In other words,

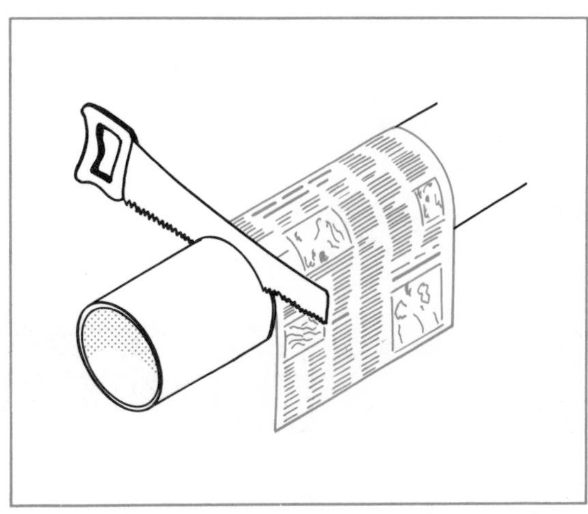

To ensure a square cut with large diameter PVC or pitch fibre pipes a sheet of newspaper may be laid over the pipe and its edges brought together underneath. This then acts as a guide. When sawing through pitch fibre pipe lubricating the saw blade with water will prevent it from sticking.

draw a pencil line round the pipe at the socket edge, then withdraw 10 mm and draw a further line. The second line will be the one to which the pipe end is finally inserted.

Clean the recess within the pipe socket and insert the ring joint. Lubricate the pipe end with a small amount of petroleum jelly (vaseline) and push the end firmly home into the socket through the joint ring. Adjust the pipe position so that

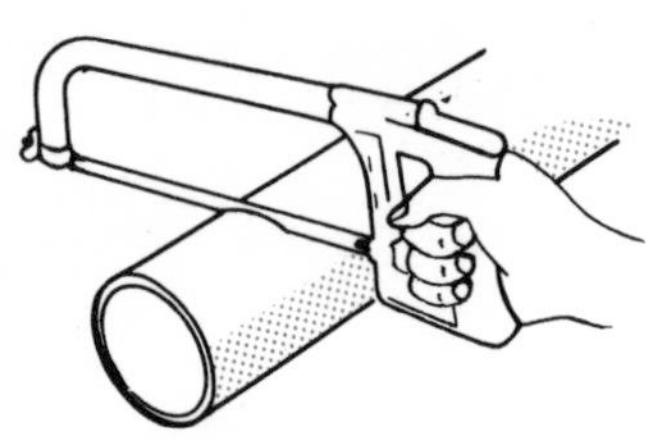

1. Cut tube squarely with fine tooth saw

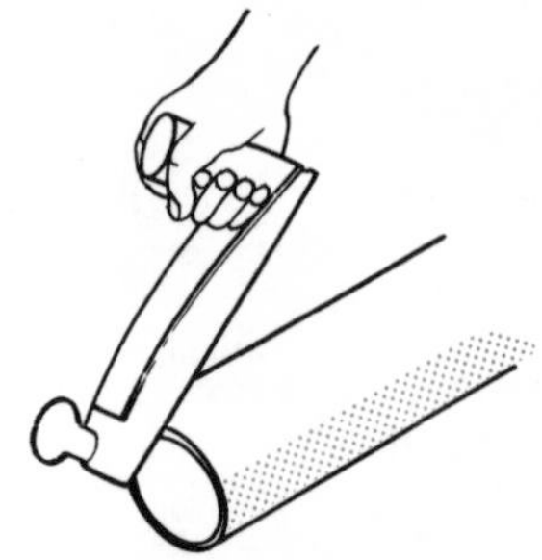

2. Chamfer tube end

3. Mark depth of tube in socket

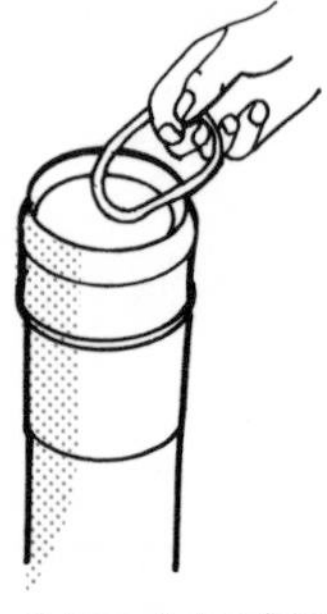

4. Insert ring joint in socket

5. Apply petroleum jelly to tube end

Ring seal in specially shaped recess

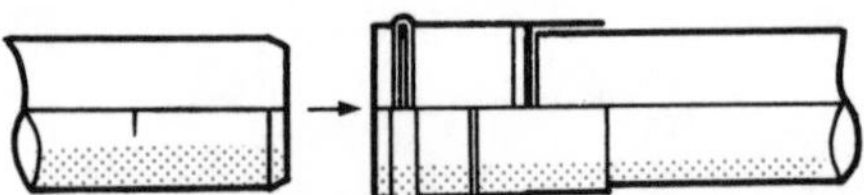

Socket solvent welded to pipe in factory on standard socket and spigot pipe length

6. Align tube end to socket and push home

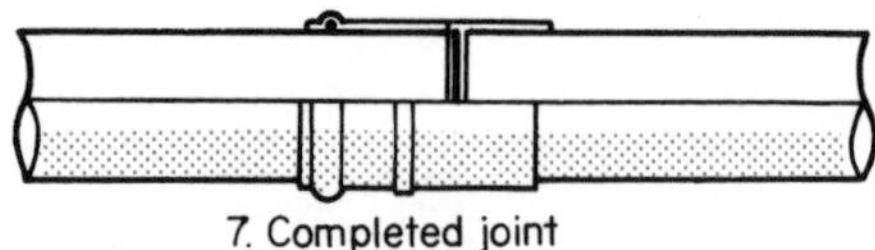

7. Completed joint

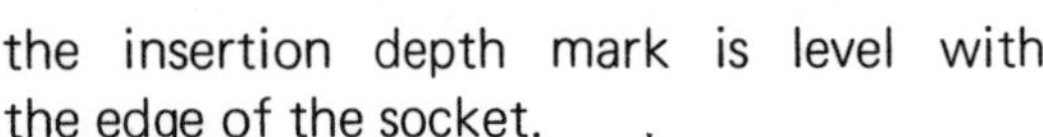

Ring seal joints are easily made as shown above and, unlike solvent welded joints, can accommodate the expansion and contraction of the pipework. Ring seal joints may be used with PVC waste and drainage systems and must be used with polypropylene systems.

the insertion depth mark is level with the edge of the socket.

A variety of means are provided by the manufacturers for connection of PVC water, waste and drain pipes to taps and ball-valves, copper or galvanised steel tubing and stoneware or iron drainage goods.

Polypropylene tubing

Polypropylene waste and above-ground drainage systems are used mainly for high temperature and chemical wastes from factories, laundries and other commercial premises.

They are not therefore very likely to be encountered by the home plumber. *The* important difference between polypropylene and PVC is that the former material cannot be solvent welded. Only ring seal joints may be used with this material.

Pitch fibre pipes

Although pitch fibre pipes are sometimes used for above ground waste and soil stacks their most common use is for underground drainage work.

There are several methods of jointing but the simplest is undoubtedly the snap ring joint. The snap ring is placed over the end of the pipe, care being taken to ensure

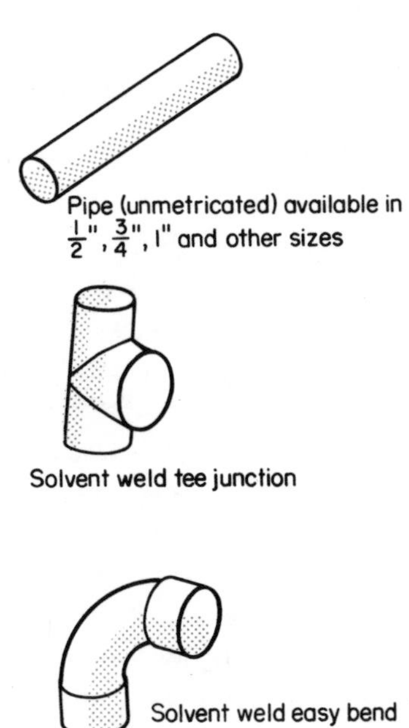

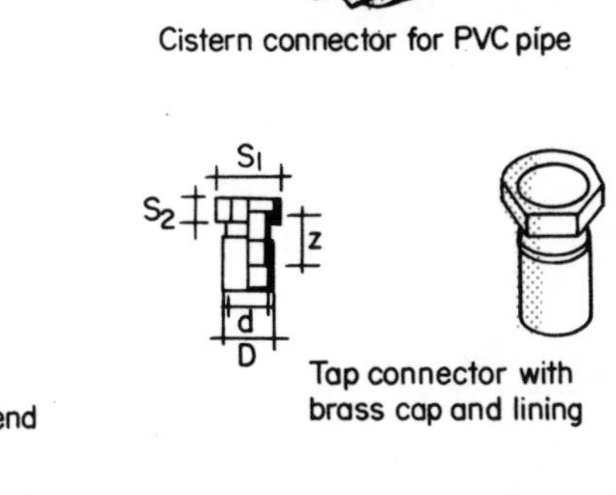

PVC cold water supply pipe and some of the solvent weld fitting available.

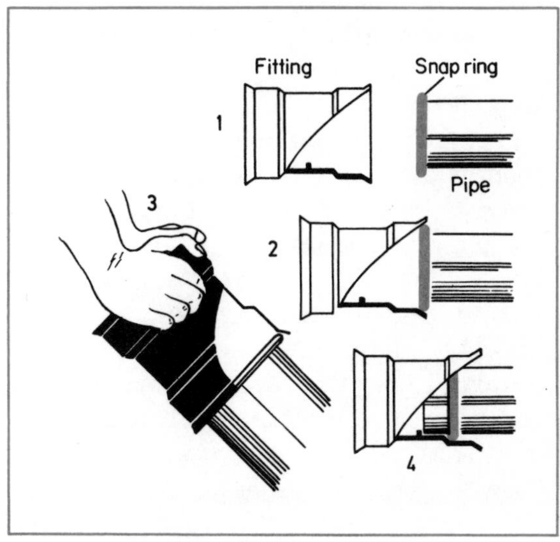

Making a snap-ring joint with pitch fibre pipe.
1 The snap ring is placed on the pipe end.
2 The joint is aligned with the tube end.
3 The joint is pressed firmly home.
4 The complete joint.

that the ring is square to the axis of the pipe and that its flat surface is in contact with the pipe. The coupling is then pushed home over the ring and pipe end so as to force the ring to roll along the pipe.

Due to the shape of the section of the ring it is compressed and jumps into its final position. This can be distinctly felt and is an indication of a sound joint.

As with PVC waste and drain pipes a variety of means are provided for the connection of pitch fibre drain pipes to pipes of other materials.

The astute reader will have noted that, in discussing PVC and pitch fibre underground drain pipes, I have omitted to give the diameters of the pipes.

This must be blamed on to the progress of metrication. Domestic drainage, in the old 'imperial' days, was always carried out in 4 in pipe. It is still carried out in the equivalent of 4 in pipe. However the makers of pitch fibre pipes described theirs as being 100 mm (the internal diameter of the pipe) while the makers of PVC pipes describe theirs as being 110mm (the external diameter of the pipe).

Appendix Some rural plumbing problems

A cottage in the country, remote both from the city's smoke and bustle and the soul-destroying monotony of suburbia, is every townsman's dream. Some realise it, either for retirement or for commuting to the office, only to find that what had appeared to be a dream was, in fact, a nightmare!

Rural life has its problems too; very different from those of the town. Foremost among them is likely to be the question of drainage. Town dwellers are so accustomed to pulling the plugs out of their baths, sinks and wash basins and operating the flush of their lavatories, that they tend to forget these facilities are possible only because a comprehensive system of street sewerage, culminating in a large-scale sewage treatment works, has been provided.

Many villages and rural communities nowadays have their own sewerage systems and sewage treatment plant. There are however still many isolated properties which have no access to a public sewer.

It does not necessarily follow that, because a village has an up to date sewerage system, every property in the village is connected to it. Local authorities' power to compel existing properties to connect to a newly laid sewer is far more limited than many people imagine.

Far too many townsmen, when buying a rural property, accept the house agent's breezy, 'There's cesspool (or septic tank) drainage of course', with only the vaguest idea of what this may mean. After a few months' residence they may be only too well aware of its implications!

Cesspools

It is not uncommon for the descriptions 'cesspool' and 'septic tank' to be used as though they meant the same thing. In fact they are vastly different. A cesspool (known in some parts of the country as a 'cesspit' or a 'dead well') is simply an underground watertight tank designed to hold sewage ***temporarily***. It must be emptied regularly and frequently.

A country cottage built between the wars or earlier might well have a cesspool of the size and type illustrated. It is 6 ft in diameter and has an effective depth of 6 ft below the level of the inlet. Its capacity in cubic feet can be ontained from the formula ($\pi r^2 h$) that you learned at school to calculate the volume of a cylinder. Multiply the square of the radius by 3.14 and multiply the result by the effective depth. In the example given the calculation would work out like this: $3^2 \times 3.14 \times 6 = 169.56$ cu.ft.

Cubic feet can be converted to gallons by multiplying by 6.25. The cesspooltherefore has a capacity of 169.56 x 6.25 = about 1 000 gal. This sounds like quite a lot of sewage. How long would the cesspool take to fill?

Waterworks authorities allow something like 45 to 50 gal of water per head of the population per day. This includes street washing and industrial purposes so it is probable that for domestic use only (including baths, clothes washing, flushing and so on) between 20 gal and 25 gal per day should be allowed for each individual. All of this water will find its way, in one form or another, into the cesspool.

Taking the lower figure of 20 gal per day, it will be obvious that a family consisting of a husband and wife only would fill

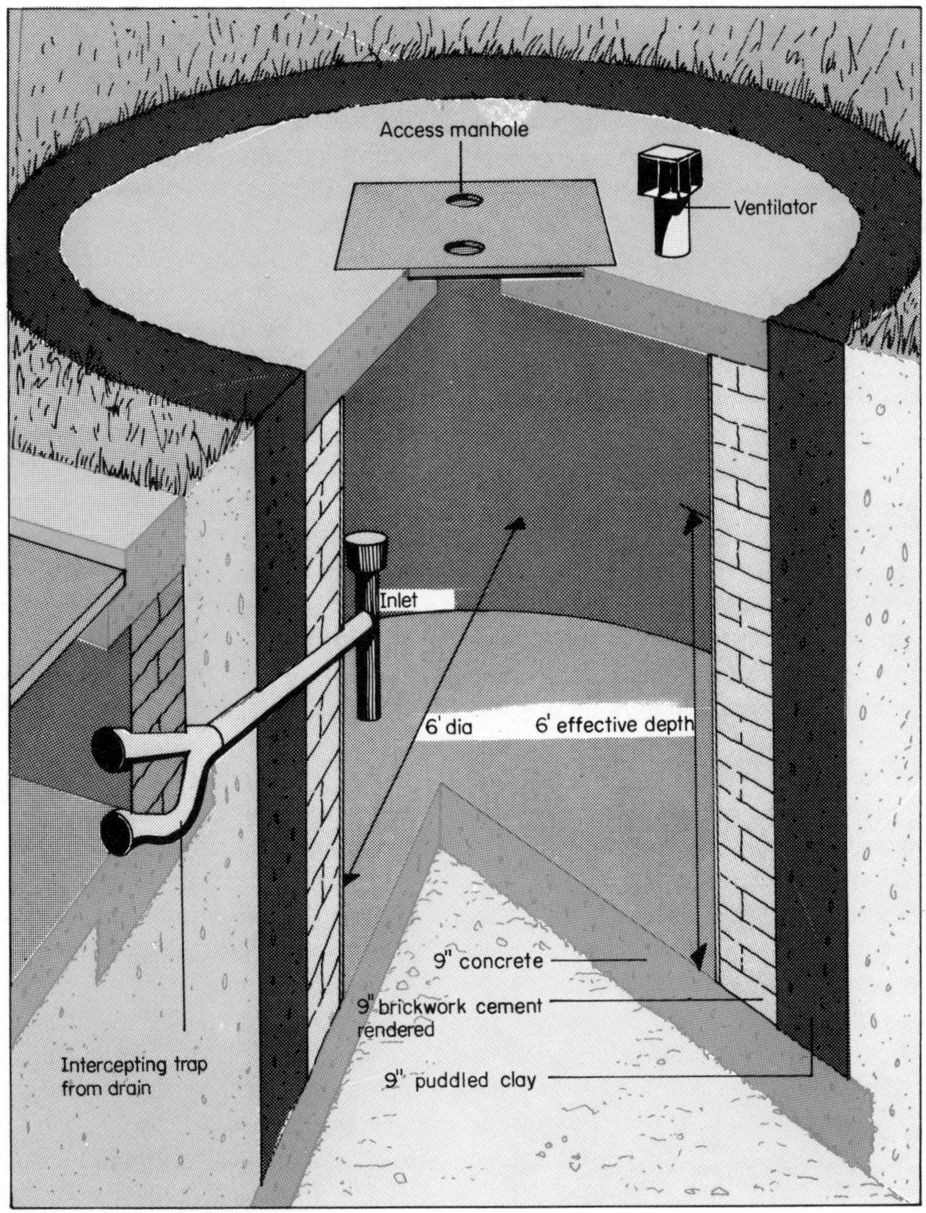

A typical small cesspool serving a pre-Building Regulation cottage. Never buy a cottage with drainage in this kind in ignorance of the cost and availability of the cesspool emptying service. You will need it very regularly!

the cesspool in just under a month. If they had two children it would be full in less than a fortnight!

The need for continual emptying—a smelly, unpleasant business under the best of circumstances—is the big snag about cesspool drainage.

Before even considering the purchase of a house with cesspool drainage check the capacity of the cesspool and work out how often it will need to be emptied. Check with the local district council about the cesspool emptying charges. Most local councils in rural areas operate a cesspool emptying service. If they don't, they will certainly be able to put you onto a private contractor who undertakes this work.

Find out about the charges for the cesspool emptying service and about its ***availability.*** If your cesspool is overflowing and the drains are backing up it is little comfort to be assured by even the most sympathetic voice on the phone at the council offices that, 'your cesspool will be emptied within a week or ten days, sir'.

Some local authorities offer a free cesspool emptying service though this may be limited to a fixed number of emptyings per year. Others operate a subsidised service and yet others expect cesspool owners to pay the full economic cost of emptying. Charging the full cost to the owner of a cesspool drained house within a sewered area is one way in which local councils may encourage connection to the sewer.

The basis on which the charge is made may also vary considerably from area to area. It may be a flat rate of so much per cesspool or so much per load on the cesspool emptier. It may, on the other hand, be an hourly charge based on the time between the cesspool emptying vehicle leaving its depot and its return there after completing the job and discharging the cesspool contents.

Prospective house purchasers, raising the question of cesspool emptying with an estate agent or owner, are sometimes answered with a knowing wink and the assurance; 'There's no need to worry about emptying ***this*** cesspool sir. There's a hole in the bottom. It'll never need emptying'.

Discount such assurances. Quite apart from the illegality—and potential danger to health—of a leaky cesspool, a moment's reflection will make it obvious that, if the contents can leak out, subsoil water can leak in. In many parts of the country, where the subsoil water level is high, this is a far more probable occurrence.

I have seen subsoil water pouring through the walls of a leaky cesspool like a miniature Niagara as its contents have been sucked out by the cesspool emptier. In a wet season such a cesspool can be filled again to overflowing before the emptier has reached its depot.

The Building Regulations recognised that the overwhelming majority of existing cesspools were far too small. They require that the minimum capacity of new cesspools must be 4 000 gal; almost four times that of the pre-Building Regulations cesspool that we discussed earlier in this chapter.

Don't forget though that this new minimum applies only to cesspools constructed after the Building Regulations came into force in the mid-'60s. For many years to come most cesspools in this country will be far smaller than that.

The new large capacity cesspool should need emptying far less frequently than its predecessor. With a family of two it should need emptying only some three times a year.

The large capacity has snags however. Older and smaller cesspools could usually be emptied with one load of the cesspool emptier. The vehicle may have to make several expensive visits to empty 4 000 gal.

Then, of course, a large cesspool is much more expensive to build, and much more difficult to make watertight, than a smaller one. A cesspool with a capacity of 4 000 gal must have a cubic capacity, below the drain inlet, of 644 cu. ft; this could be obtained with a cesspool 12 ft deep and with an area of 54 sq. ft.

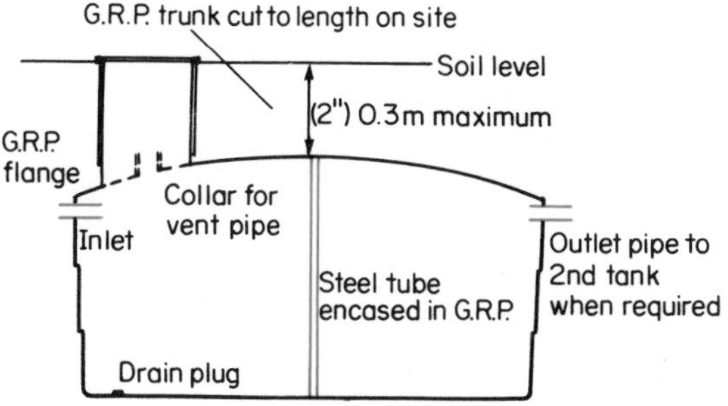

This 2,000 gal capacity plastic/glass fibre cesspool made by Rokcrete Ltd. will be watertight and need less frequent emptying. Remember though, that all forms of cesspool drainage only postpone disposal of the sewage.

It should be added that some local authorities will modify the requirements of the Building Regulations under certain special circumstances. They may, for example, accept a 2 000 gal capacity cesspool in an area expected to be sewered within, say a year.

Recognising the difficulty of constructing an absolutely watertight large capacity cesspool an Essex firm has produced a glass fibre reinforced plastic cesspool of 2 000 gal capacity which, being made in one piece, can be guaranteed to be watertight. Two of these can be linked together to give a capacity of 4 000 gal. They have been successfully installed in rural areas of north-east Essex, even where there has been an exceptionally high water table in conjunction with wet running sand.

Septic tanks

So much for cesspools. A septic tank is, or should be, a different matter entirely. At its best, a septic tank system is a small, private sewage treatment plant which will function satisfactorily, with the minimum of attention, for years.

The septic tank itself is an underground chamber designed to retain sewage for at least 24 hours. During this period the sewage is liquefied by the action of bacteria. A thick scum forms on the surface of the liquid in the tank and sludge forms at the base. The liquid in the middle is drained off from beneath the scum.

A typical small septic tank system, designed by Burn Bros. (London) Ltd, is illustrated. Sewage enters, without disturbing the surface of the liquid already in the tank, by means of a dip-pipe inlet. It leaves

A septic tank and filter installation designed by Burn Bros. Ltd. A properly designed septic tank installation can provide a satisfactory sewage treatment home for a single house or small group of houses. Always check with the District or Borough Council before committing yourself to any expense in connection with the installation of such a plant.

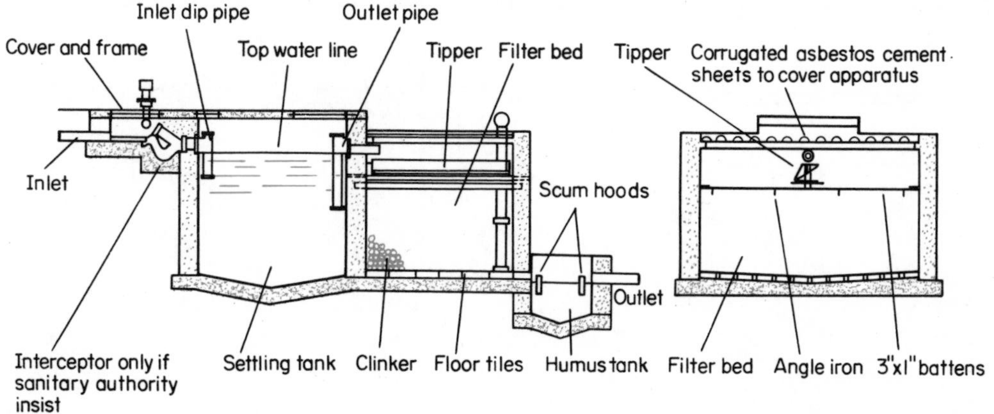

by a similar dip-pipe at the other end of the tank.

Septic tanks do not purify sewage. They simply liquefy it. Before it can be discharged into a ditch or stream it is necessary to aerate it by allowing it to percolate slowly through a 3 ft to 4 ft deep bed of stone or clinker.

Aeration, like septic tank action, is a bacterial process. The effect of septic tank treatment and aeration is to speed up and control the natural processes of decomposition. The complicated chemical constituents of untreated sewage are broken down by bacterial action into harmless and inoffensive nitrites and nitrates dissolved in the final effluent.

The bed of stones or clinker through which the effluent from the septic tank passes is usually referred to as a filter bed. It must be stressed however that its function is aeration and ***not*** filtration.

On no account should what is sometimes known as a 'submerged filter' be constructed. This kind of filter is sometimes encountered on a level site where there is not sufficient fall for the construction of a proper filter bed. It has its inlet and outlet at the same level. The filtering medium is therefore constantly submerged and is consequently quite unable to perform its sole function—that of aeration.

If there is insufficient fall for a conventional filter bed with a gravity outfall, then a collecting pit should be constructed below the level of the filter outlet. This should be provided with a float operated electric pump to raise the treated final effluent to its outfall level.

It will be seen that the septic tank and filter system illustrated has an automatic tipper which ensures that the effluent from the septic tank is distributed first over one side of the filter medium and then over the other. This ensures even distribution and allows each side of the filter bed to have a 'recovery period'.

The manufacturers of the plastic/glass fibre cesspool have also produced a ready made 700 gal capacity septic tank unit in the same material. This can be installed as quickly and easily as the cesspool but

(a) Ready made plastic/glass fibre septic tank manufactured by Rokcrete Ltd. The effluent will need further treatment either by aeration through a 'filter bed' or by subsoil irrigation.

(b) Precast concrete septic tank and filter manufactured in sections by Ingol Precast Ltd. This amounts to a compact 'packaged' sewage treatment plant. Septic action takes place in the inner chamber and the effluent overflows from the drip edge to percolate through an aerating filter.

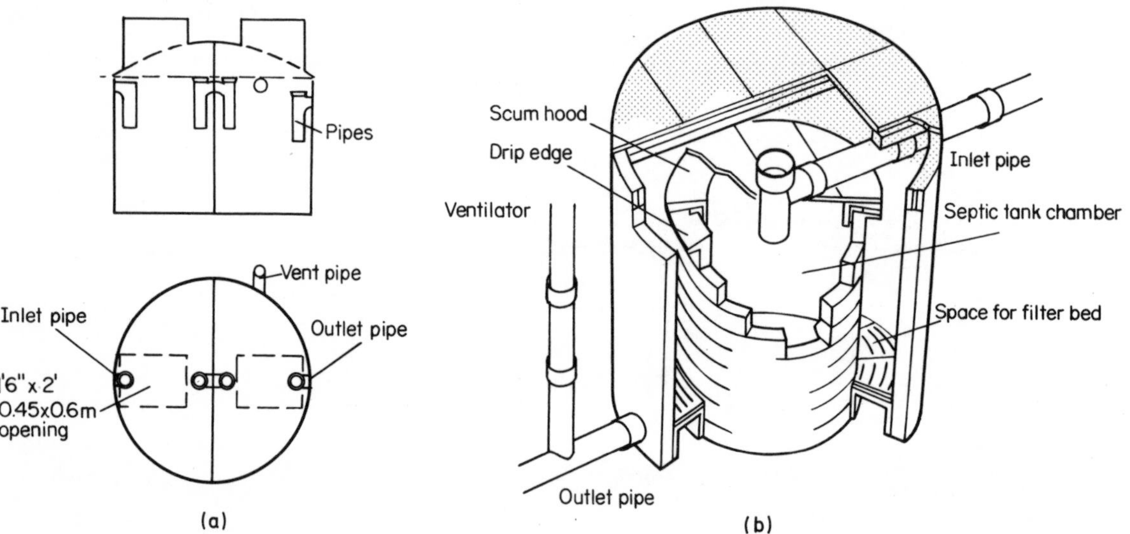

it will, of course, normally need the provision of a filter.

'Packaged sewage works'

A 'packaged sewage works' incorporating both septic tank and filter is made by Ingol (Precast) Ltd. of Preston, Lancs. The cutaway diagram shows the method of construction and principle of operation.

Sewage enters via the inlet pipe and discharges into the circular septic tank through the central dip pipe. The scumhood prevents the scum which forms on the surface from washing out on to the filter medium.

The effluent flows under the scum hood and on to the surface of the circular filter bed from the srip edge. After passing through the filter the effluent is collected at the base and passes out of the system to its outfall.

Drainage

Under very favourable circumstances—an absorbent subsoil, no sources of water in the vicinity and a sufficient area of land—it may sometimes be found possible to dispense with the filter or aerator and to dispose of the effluent from the septic tank by subsoil irrigation.

Land drains for this purpose should be laid flat, or almost flat, on a 1 ft deep bed of clinker or brick-bats. There should be a further 1 ft of this material on each side of the pipeline. Perforated pitch fibre land drain pipes are the best for this purpose.

Some 100 ft to 150 ft length of pipe will be necessary in average soil. Depth will depend upon the depth of the outlet from the septic tank but it should be kept as shallow as possible. The soil near the surface is generally the more absorbent.

With a system of this kind it is usual for the effluent from the septic tank to flow directly into the land drainage system. A disadvantage of this is that the soil in the immediate vicinity of the septic tank outlet becomes very heavily charged with sewage while for many weeks or months no effluent will reach the far end of the pipe line.

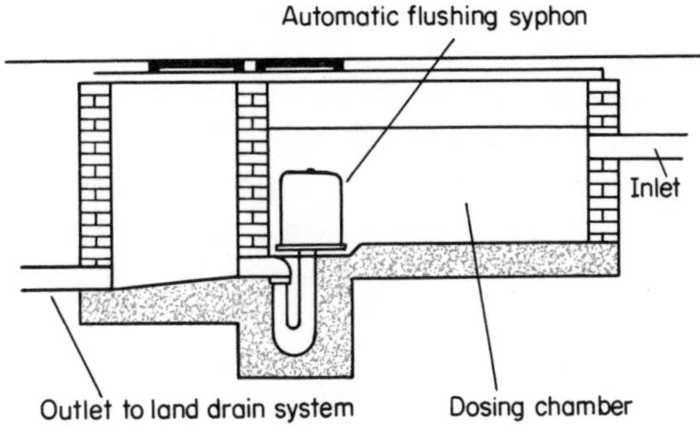

Where it is possible to dispose of septic tank effluent by subsoil irrigation the provision of a dosing chamber and automatic flushing siphon will ensure that the effluent is distributed evenly throughout the system and that the subsoil has an opportunity to 'recover' from each dosing.

To avoid this, it is a good idea to construct a final dosing chamber at the point where the effluent is discharged from the septic tank and to equip it with an automatic flushing siphon. With this arrangement the level of effluent will rise in the dosing chamber until it reaches the point at which the automatic flushing siphon comes into operation, All the liquid in the dosing chamber will then be released in one flush throughout the land drainage system.

This ensures that the effluent is distributed evenly through the system and that the soil in the immediate vicinity of the septic tank is not overloaded and soured. It also ensures that, after each flush, there is a period of time in which the soil can absorb, and the soil bacteria act upon, the released effluent.

Maintenance

A properly designed system of this kind should need little maintenance beyond a periodic clearance of sludge from the bottom of the tank.

Rain water should always be excluded both from cesspools and from septic tanks. Discharge rain water drains directly into a ditch or construct a soakaway.

Since septic action is bacterial action, those who have a septic tank system should not use strong disinfectants excessively. No harm will result from using a household disinfectant or drain cleanser on the gullies and lavatories once a week.

Excessive use of detergents is also unwise. Some housewives do tend to be overgenerous in their use of these materials. If overused, synthetic detergents can emulsify the fats naturally present in sewage. Instead of a scum forming on the top of the septic tank and a sludge at the bottom with a relatively clear fluid between, a liquid of soup-like consistency will fill the tank. This will wash through the outlet to clog the filter or land drainage system.

Taking precautions

If you are thinking of purchasing a rural cottage which is not connected to a sewerage system examine its drains and their outfall in the light of this chapter. If there is a cesspool, check its capacity and the cost and availability of the cesspool emptying service. If there is a septic tank system check on its eventual outfall; into a ditch or stream perhaps. If it is black and evil smelling you can be fairly certain that the time is not far distant when the council will require that state of affairs to be remedied.

It may be that your rural home has not any drainage system at all. Waste water is thrown on to the garden and there is an old fashioned pail closet. Part of your modernisation scheme will be to provide a comprehensive drainage system, bath, sink, toilet and all the trappings of modern civilisation.

In this event you should call at the district council offices and have a chat with the building control officer, or the environmental health officer, before committing yourself in any way. It may be that some kind of septic tank installation is possible. Here you must take the advice of the local official, A basic septic tank system that could be perfectly satisfactory for a remote moorland cottage could be a serious threat to health on the outskirts of a populous village.

Think twice, and then again, about the provision of a cesspool. Remember the responsibilities that it entails and the fact that it only ***postpones*** the problem of disposing of the household's sewage. I personally can imagine no circumstances in which I would be prepared to buy a home with a cesspool or one where it was necessary to construct one.

Find out if, or when, a public sewer is likely to be constructed near your proposed new home and make your plans accordingly, If this is likely to be within say, a year, you might well be wise to postpone the installation of your drainage system.

The idea of a chemical closet may lack appeal to the townsman for permanent use. Yet modern chemical closets (Elsan, Racasan and Perdisan) are very different from the old pail or earth closet. To my mind they are infinitely preferable to having a modern flush lavatory that one is afraid to use because the cesspool is already overflowing–and with no hope of its being emptied for days!

Water from a well

Water supply is less likely than drainage to be a problem to the seeker after rural solitude. Local authorities have, in the past, always inclined to provide mains water services in advance of any sewerage system and there is no reason to suppose that the Area Water Authorities, who are now responsible for both these services, will adopt any different policy.

There are, of course, still many rural cottages supplied with water from a well. The prospective purchaser would be wise to regard them with suspicion; particularly if the well is of the picturesque brick type beloved by the producers of picture post-cards.

The 'olde worlde' thatched cottage of these postcards traditionally had two holes in the back garden: the well and the cess-pool. Not even the most fanatical con-servationist is likely to feel too much enthusiasm for the kind of 'recycling' produced by this arrangement.

Discount assurances that the previous inhabitant of the cottage was the healthiest man in the village and that he was accident-ally killed while hunting at the age of 104. He may well have had fifteen children, all of whom died in infancy from typhoid fever!

Never rely on a water supply of this kind without first asking the Environ-mental Health Officer of the local District Council to sample the water for bacterio-logical analysis. It would take several negative samples taken over a considerable period of time before I felt happy about it.

If you have, or there is the possibility of your having, a young baby, it would be wise to arrange for the well to be sampled for chemical as well as bacteriological analysis. Water that is bacteriologically satisfactory may contain excessive nitrites which, if used in a baby's feed, can extract the oxygen from his blood to produce the 'blue baby' condition.

Local Councils normally take routine chemical samples from wells serving homes in which there is an expectant mother. If the water has excessive nitrites they will make arrangements for an alternative water supply for use until the child is old enough to be able to cope with them.

There are, in fact, two kinds of well: deep and shallow. ***Actual*** depth, from ground level to water line or from ground level to the base of the well, is not the deciding factor in classification.

A 'shallow well' is one that depends for its water supply upon subsoil water—water that is found ***above*** the first impermeable rock stratum beneath the soil. On a good building site, beneath a layer of top-soil 1 ft to 1 ft 6 in deep, there may be a 15 ft depth of gravel lying on a 20 ft thick bed of impermeable clay.

A well, dug on such a site, drawing water from above the clay would be a shallow well. If, on the other hand, the well were dug deeper—through the clay stratum to a water-bearing stratum, perhaps chalk, under-neath it would be described as a deep well.

A shallow well taps subsoil water held above the first impervious stratum. Water from such a well is suspect. A deep, and generally safe, well taps water held below the first impervious stratum. The actual depth of the well has no bearing on the classification.

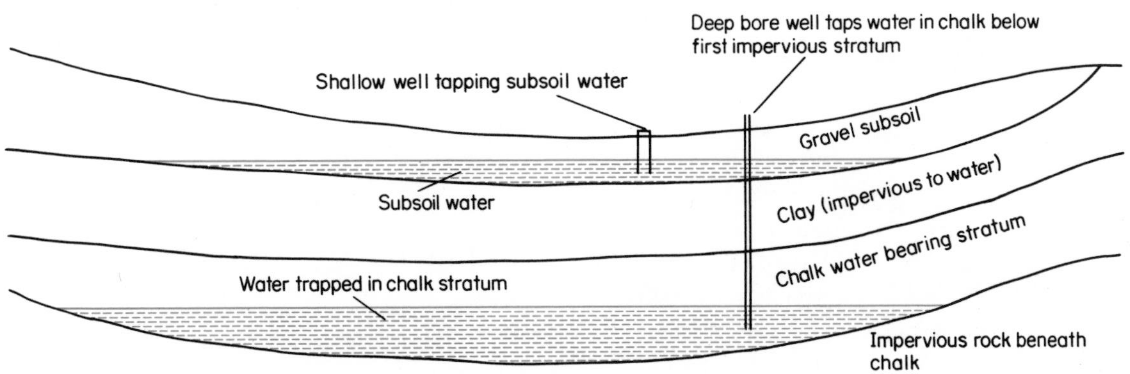

Water from a deep well is much more likely to be satisfactory for drinking purposes because of the additional layers of rock through which the water has filtered to its underground reservoir.

Dug, brick-lined wells are generally shallow ones. Deep wells are usually bored and are often referred to as 'bore wells'.

A bore well of this kind will probably provide you with a safe and wholesome water supply. It is still worth while though to arrange for the Environmental Health Officer to check it for you.

If you have a water supply of this kind you will probably have a float activated electric pump to pump it up to the main storage cistern in the roof space.

Don't overlook the fact that this will mean that ***all*** your domestic water, including that used for cooking and drinking, will come from the roof storage cistern. This makes it imperative that the cistern should be kept clean and that it should have a cover capable of excluding the mice, birds (and possibly, bats) that may find themselves in a rural roof space.

Index